# Ancient
# Sedimentary
# Environments

# Ancient Sedimentary Environments

and their sub-surface diagnosis

## THIRD EDITION

RICHARD C. SELLEY

Cornell University Press

Ithaca, New York

First published 1970
Second edition, revised 1978
Third edition, revised 1985


Library of Congress Cataloging-in-Publication
Data

Selley, Richard C., 1939–
    Ancient sedimentary environments and their
    sub-surface diagnosis


    Includes bibliographies and indexes.
    1.  Rocks, Sedimentary.      2.  Geology,
    Stratigraphic.
1. Title.
QE471.S42      1985          551.3′03          85-17424
ISBN 0-8014-1871-2
ISBN 0-8014-9354-4 (pbk.)



Printed in Great Britain

# Contents

# Preface to the first edition

In this book I have attempted to show how the depositional environments of sedimentary rocks can be recognized. This is not a work for the specialist sedimentologist, but an introductory survey for readers with a basic knowledge of geology.

Within the last few years environmental analysis of ancient sediments has been enhanced by intensive studies of their modern counterparts. Thus Geikie's dictum 'the present is the key to the past' can now be applied with increasing accuracy. While an understanding of Recent processes and environments is critical to the interpretation of their ancient analogues, it is beyond the scope of this book to describe these studies in detail. I have, however, attempted to summarize the results of this work; detail being sacrificed in the interest of brevity. Inevitably, this has led to a tendency to generalize; I have tried to counteract this by providing bibliographies of studies describing Recent sediments.

The economic importance of environmental analysis of ancient sediments is increasing. With the world expected to consume as much oil and gas in the decade 1970–80 as has been produced to date, the search extends increasingly to the more elusive stratigraphically controlled accumulations.

Similarly environmental analysis has a part to play in locating metallic ores in sediments whose geometries are facies controlled.

The book begins with a discussion of the classification of sedimentary environments and an evaluation of the methods which may be used to identify them in ancient deposits.

Each subsequent chapter describes a particular depositional environment, beginning with a summary of its characteristics as seen on the earth's surface at the present time. This is followed by a description of an ancient case history whose origin is then deduced. A general discussion of the problems of identifying the environment in ancient sediments comes next, and each chapter concludes with a brief review of its economic significance.

Neither the selection of environments nor the discussion of economic aspects is intended to be comprehensive. I hope, however, that there are sufficient examples to show how the environment of a sedimentary rock can be determined and some of the ways in which sedimentology can be used in the search for economic materials. The discussions of the economic aspects of the various environments are heavily biased in favour of the oil industry at the expense of mining, hydrogeology, and engineering geology. This is no accident. The oil industry is the largest employer of sedimentologically oriented geologists and has done more to advance and apply sedimentology than any other branch of industrial geology.

Critical readers will notice that metres, feet, kilometres, and miles are used indiscriminately throughout the book. Since the oil industry refuses to go metric, the student must quickly learn to correlate the two systems. A conversion scale is included in the first figure.

*January, 1970*                                                    RICHARD C. SELLEY
*Tripoli, Libya*

# Preface to the second edition

I began to write the first edition of this book after spending nearly ten years at university studying and teaching how to diagnose the depositional environments of sedimentary rocks where they crop out at the earth's surface.

The first edition was written during the first three months of my five-year sabbatical in the oil industry. From then on a very large part of my time has been spent learning how to diagnose the environments of sediments from bore holes in the sub-surface. This is a far more challenging occupation for there are fewer data, the techniques are quite different, and the economic implications may be immense.

The new edition reflects this experience. The introductory chapter includes a discussion of the techniques of sub-surface facies analysis, and subsequent chapters discuss the criteria by which each environment may be recognized in the sub-surface.

Most of the chapters have been modified in one way or another; sections on modern environments have been expanded and some of the case histories have been extensively modified and, for the Captain 'reef', completely rewritten.

I have updated the bibliographies, but since the sedimentology part of GeoAbstracts show that some 20 000 papers have been published in this field since the first edition, I may well have missed one or two important papers. I apologize to the reviewers, for experience has taught me that the unquoted seminal papers are inevitably their own.

*August, 1977*                                                      RICHARD C. SELLEY
*Imperial College, London*

# Preface to the third edition

I wrote the first edition of this book at a very impressionable age. I had just completed nine years of university research on the environmental interpretation of sediments, and had started work for an oil company. The first edition was thus concerned with the interpretation of sedimentary environments from outcrop, from an aesthetic stance, with no thought of vulgar commercial applications.

I wrote the second edition after spending several years learning how to interpret the depositional environment of sediments in the subsurface, and in trying to apply this knowledge to petroleum exploration. Thus the second edition was expanded to include sections on the use of geophysical well logs in subsurface facies analysis.

Within the last few years seismic geophysical surveying has improved tremendously. A whole new field has developed, loosely termed seismic stratigraphy, which applies sedimentary concepts to the interpretation of seismic data. Today seismic surveys may delineate channels, deltas, reefs, submarine fans, and other deposits. In this edition I have included sections on the seismic characteristics of the various sedimentary facies. I have also expanded sections on metalliferous sedimentary deposits, and on recent environments. New case histories have been introduced and I have attempted the impossible task of updating the bibliographies.

*September, 1984*                                    RICHARD C. SELLEY
*Imperial College, London*

# Acknowledgements

Any textbook, by its very nature, is parasitic on previously published work. This is probably more true of this book than many others due to its case history approach.

I am, therefore, extremely grateful to the authors of the various case histories who allowed their work to be pirated in this fashion, who criticized portions of the manuscript, and who generously supplied photographs. These include J. R. L. Allen, A. H. Bouma, J. D. Collinson, A. Hallam, A. J. Jenik, J. F. Lerbekmo, N. D. Newell, M. D. Picard, H. G. Reading, D. J. Stanley, W. F. Tanner, G. S. Visher, R. G. Walker, and R. J. Weimer.

Correct opinions and environmental interpretations are attributable to the authors of the case histories. Errors of fact and interpretation are due to me.

My colleagues at the Oasis Geological Laboratory, particularly Dr J. Hea and D. Baird, suggested numerous improvements to the manuscript. Thanks are also due to Oasis Oil Company of Libya Inc. for permission to publish.

I am extremely grateful to Professor J. Sutton and the Imperial College authorities who granted me leave of absence to retire from the academic hurly-burly to the peace and quiet of the oil industry to write this book.

R.C.S.

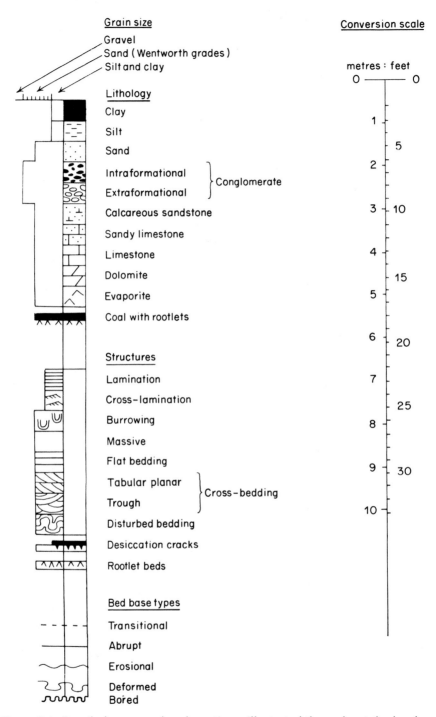

**Grain size**

Gravel
Sand (Wentworth grades)
Silt and clay

**Conversion scale**

metres : feet

**Lithology**

Clay

Silt

Sand

Intraformational

Extraformational

} Conglomerate

Calcareous sandstone

Sandy limestone

Limestone

Dolomite

Evaporite

Coal with rootlets

**Structures**

Lamination

Cross-lamination

Burrowing

Massive

Flat bedding

Tabular planar

Trough

} Cross-bedding

Disturbed bedding

Desiccation cracks

Rootlet beds

**Bed base types**

Transitional

Abrupt

Erosional

Deformed
Bored

*Figure 0.1* Detailed measured rock sections, illustrated throughout the book, are drawn using the above key. Note that grain size is drawn increasing to the left so as to correlate with wireline logs.

# 1 Introduction

## SEDIMENTARY ENVIRONMENTS AND FACIES

*A sedimentary environment* is a part of the earth's surface which is physically, chemically, and biologically distinct from adjacent terrains. Examples include deserts, river valleys, and deltas.

The three defining parameters listed above include the fauna and flora of the environment, its geology, geomorphology, climate, weather, and, if subaqueous, the depth, temperature, salinity, and current system of the water. These variables are tightly knit in dynamic equilibrium with one another like the threads of a spider's web. A change in one variable causes changes in all the others.

Numerous deserts, lakes, deltas, reefs, and other environments are found all over the present-day face of the earth suggesting that there are a finite number of sedimentary environments. This statement must be qualified though by noting that no two similar environments are ever exactly alike, and that different environments often merge imperceptibly with one another across the face of the earth.

A sedimentary environment may be a site of erosion, non-deposition, or deposition. As a broad generalization, sub-aerial environments are typically erosional while sub-aqueous environments are mostly depositional areas. Some environments alternate through time between phases of erosion, equilibrium, and deposition. River valleys are a case in point.

Erosional environments generally occur in dissected mountain chains, and to a lesser extent along rocky shores. They are not preserved in the statigraphic record and we can only infer their past existence from adjacent detrital formations.

Equilibrial or nondepositional environments occur both on land and under the sea. Many deserts are equilibrial surfaces on which there is neither erosion nor net deposition. The wind may transport sand in eolian dunes, but these bed forms often migrate across scoured bedrock or gravel surfaces. Similarly many continental shelves are equilibrial surfaces where there is neither erosion nor net deposition taking place, though as with deserts, tidal sand waves may migrate across scoured rock or lag gravel sea floors. Equilibrial environments also occur in deep oceanic environments. Here the sedimentation rates may be so slow that the rate of solution of sediment particles may keep pace with deposition. Equilibrial environments, like erosional ones, do not actually deposit detritus. They are commonly preserved in the stratigraphic record as unconformities. Because the rock beneath an equilibrial surface is generally very well cemented they often generate velocity contrasts with the overlying strata and thus may show up as reflecting surfaces on seismic surveys.

The third type of sedimentary environment is the depositional one. It is the depositional sedimentary environment which is actually preserved in the stratigraphic record in the form of sedimentary facies.

A *sedimentary facies* is a mass of sedimentary rock which can be defined and distinguished from others by its geometry, lithology, sedimentary structures, palaeocurrent pattern, and fossils.

The consensus of geological opinion is that there are a finite number of sedimentary facies which occur repeatedly in rocks of different ages all over the world. Comparison with Recent sediments suggests that these can be related to present-day depositional environments. As with their Recent counterparts, therefore, no two similar sedimentary facies are ever identical, and gradational transitions between facies are common.

A sedimentary facies is the product of a depositional environment, a special kind of sedimentary environment. One of the main problems of determining the origin of ancient sediments is that, though essentially reflecting depositional environments, they also inherit features of earlier erosional and non-depositional phases. Consider for example an old river channel. The profile of the stream bed reflects an erosional environment, the infilling sediment reflects the nature of the source rocks and the hydraulics of the current which transported (as well as deposited) the sediment, while rolled bones and wood are derived from non-depositional environments outside the channel. It is only the sedimentary structures (and palaeocurrents) which unequivocally indicate the depositional environment.

These concepts of sedimentary environments and facies are summarized in Table 1.1. A more detailed review of the concepts of environments and facies, and their historical evolution, will be found in Selley (1982, pp. 255–266).

Table 1.1. The relationship between sedimentary environments and sedimentary facies.

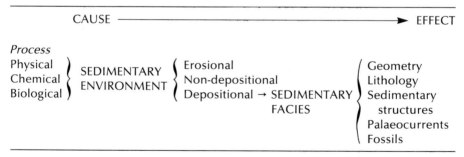

It is very important to distinguish a sedimentary environment from a sedimentary facies. There is no problem in identifying the environment of Recent sediments. A sand sample collected from a beach today is, by definition, a beach sand. When studying ancient sediments, however, it is best to classify them into facies on a purely descriptive basis; it is unwise to give them environmental names. Thus one should talk of pebbly channel sand facies, flysch facies, and so on, rather than of fluviatile facies or turbidite facies.

Many attempts have been made to classify both Recent sedimentary environments and ancient sedimentary facies (e.g. Pettijohn, 1957, p. 610; Krumbein and Sloss, 1959, p. 196; Visher, 1965; Potter, 1967; Shelton, 1967; and Crosby, 1972).

The classification of depositional environments used in this book is unique, but owes much to previous ones (Table 1.2). It is put forward not as a final statement of environments but as a foundation on which to base the case histories discussed in subsequent chapters.

*Table 1.2. A classification of depositional environments. Like all classifications this one contains several deficiencies and inconsistencies which are inherent due to the complexity of environments. Note particularly that eolian deposits can form on the crests of barrier islands: deltas can build into lakes as well as seas; reefs occur in fresh, as well as sea water.*

| | | |
|---|---|---|
| Continental | { Fluviatile<br>{ Lacustrine<br>{ Eolian | { Braided<br>{ Meandering |
| Transitional (Shorelines) | { Lobate (deltas)<br>{ Linear | { Terrigenous<br>{ Mixed carbonate: terrigenous<br>{ Carbonate |
| Marine | { Reef<br>{ Shelf<br>{ Submarine channel and fan<br>{ Pelagic | |

Recent sedimentary environments can generally be divided into sub-environments. For example, a linear shoreline is often composed of a complex of barrier islands, lagoons, and tidal flats lying between alluvial and shelf major environments. Ancient sedimentary facies can be divided in a similar manner into sub-facies which can often be attributed to sub-environments.

## RELATIONSHIP BETWEEN FACIES, SEQUENCES AND STRATIGRAPHY

Sedimentary environments occur side by side across the earth in a predictable manner. Thus for example, an alluvial flood plain may merge into a tidal flat, which may pass laterally via a lagoon into a barrier island and so to the open sea. As the sea level rises and falls the shoreline may transgress and regress the continental environment. The result of this process may be the deposition of a series of conformable facies with gradational vertical transitions. This important relationship between facies and environments was first noted by Walther (1894), and is now known as Walther's Law. This may be stated concisely as: 'A conformable vertical sequence of facies was generated by a lateral sequence of environments' (Middleton, 1973).

This concept was developed by Busch (1971) who defined the terms genetic increment and genetic sequence. A genetic increment is 'a mass of sedimentary rock in which the facies or sub-facies are genetically related to one another'. A typical genetic increment may thus consist of a single prograding delta sequence with its constituent prodelta, delta slope and delta front deposits. A genetic sequence is a conformable series of the same genetic type (Fig. 1.1).

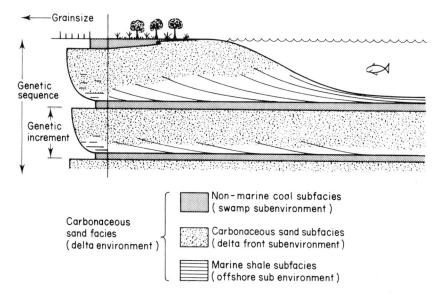

*Figure 1.1* Geophantasmogram to show the relationship between sedimentary environments and facies, to demonstrate Walther's Law, and to show how genetic increments and genetic sequences develop.

These definitions are a formal way of stating that deposition is often cyclic. One may paraphrase George Orwell and say that 'All sedimentation is cyclic, but some is more cyclic than others'. This phenomenon of cyclic sedimentation has been described and discussed in an extensive literature reviewed by Duff, Hallam, and Walton (1967).

Because of the complex interplay of the processes which control the deposition of a sedimentary sequence it is seldom that a cyclic motif, if present, is conspicuous. Instead of seeing four sub-facies arranged ABCD, ABCD, ABCD, etc., it is more usual to find something like ABCABDCDAB and so on. It is possible to deduce from the latter sequence that the 'ideal' cycle is ABCD, but this is based on a highly subjective mental process. Within recent years the eyeball method of analysing cyclic sequences has been replaced by statistical techniques.

These include simple methods which, though statistically naïve, can easily be used and understood by the field geologist (e.g. Selley, 1969b). Other, more sophisticated techniques necessitate the use of a computer, but suffer from the assumption which they have to impose on the geological data.

The advent of the computer now enables all sorts of statistical gymnastics to be applied to the study of cyclic sedimentation, both from real rock sections and from sequences generated artificially by computer modelling (Merriam, 1967; Schwarzacher, 1975).

Once the cyclic pattern has been extracted from a sedimentary sequence one is faced with the problem of its interpretation.

Cycle-generating processes have been classified into two groups defined by Beerbower (1964, p. 32) as follows:

> (*i*) *Autocyclic mechanisms:* 'are generated in the depositional prism and include such items as channel migration, channel diversion, and bar migration.'
>
> (*ii*) *Allocyclic mechanisms:* 'result from changes external to the sedimentary unit such as uplift, subsidence, climatic variation, or eustatic change.'

One of the most interesting results of studies of Recent sediments is that they have shown how sub-environments migrate laterally over one another across the depositional area resulting in a regular sequence of sub-facies. This has led to an increased understanding of the importance of autocyclic mechanisms in generating ancient cyclic sequences. Previously stress was laid on allocyclic processes.

Recognition of sequences of autocyclic origin within a sedimentary facies is an important way of identifying its depositional environment. Figure 1.2 idealizes the various sequences of autocyclic origin which may be present in different facies. It is interesting to note that these can be detected purely on a basis of vertical variations in grain size without recourse to other facies parameters. This approach can be particularly fruitful in sub-surface studies as will shortly be discussed.

## METHODS OF ENVIRONMENTAL DIAGNOSIS

There are many different techniques which can be used to determine the depositional environment of a sedimentary rock. These vary considerably according to whether the study is based on surface or sub-surface information. An alternative arrangement for this book would have been to describe and discuss methods of environmental recognition, rather than to actually show them in operation solving specific case histories. Though the latter approach seems more interesting it is appropriate first to discuss the various techniques which will shortly be employed.

As a general rule the diagnosis of a sedimentary environment should be based on a critical evaluation of all available lines of evidence. Commonly however

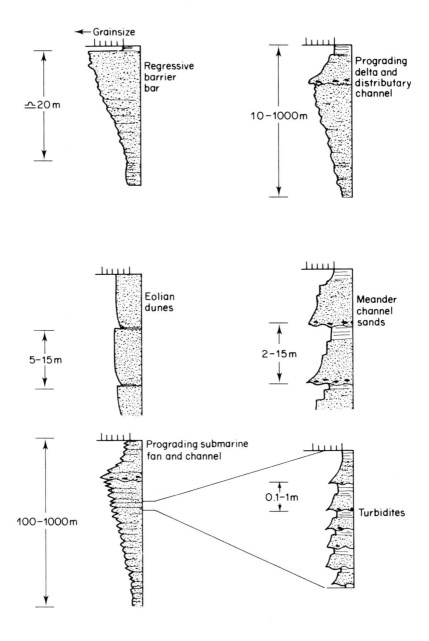

*Figure 1.2* Vertical grainsize profiles produced by some sedimentary environments. Further details and explanations are given at appropriate places in the text. The thicknesses of the various genetic increments are rough approximations only.

not all the data that are desirable are actually available. This is particularly true of sub-surface studies in which the main sources of data are seismic sections and geophysical well logs with minimal rock samples from cores and cuttings. In such cases one can only make the best use of what facts can be ascertained. In other cases, such as outcrop studies, though there may be a wealth of facts to utilize, only one or two may actually be critical to the diagnosis of the environment in question. In different cases different lines of evidence will be of varying importance. Thus the recognition of a reef may be confidently based on only lithological and palaeontological data. A delta on the other hand may be identified by its geometry and vertical sequence of grain size and sedimentary structures. These points will become clearer as the book proceeds.

The techniques of environmental analysis can most conveniently be discussed under the five defining parameters of a facies: geometry, lithology, sedimentary structures, palaeocurrent patterns, and fossils.

The chapter concludes with a discussion of the use of wireline logs in sub-surface environmental interpretation.

## Geometry

The over-all shape of a sedimentary facies is a function of predepositional topography, the geomorphology of the depositional environment and its post-depositional history. For example the geometry of blanket sands burying old land surfaces will largely be controlled by the geomorphic stage and process which previously operated on that terrain. This may give little indication of the origin of the overlying sand. Likewise post-depositional erosion and tectonic deformation will modify the geometry of a facies and inhibit the use of its over-all shape as a diagnostic feature. It is only when the syn-depositional shape of an environment is preserved that geometry is a valuable diagnostic criterion. Figure 1.3 illustrates some of the more important facies shapes that may be recognizable. It is important to note that the same geometry may be produced in one of several different environments. Channels may be fluvial, deltaic, tidal or submarine. Fans may be alluvial, deltaic or deep marine. A blanket geometry is obviously of no environmental significance. Detailed study may show however that a sheet is composed of a series of coallesced channels, shoestrings or other smaller bodies.

The geometry of a sedimentary facies may be relatively easy to determine where it crops out at the surface and where exposure is good. In the subsurface however this is impossible. Traditionally the approach was to map the geometry of facies from borehole data; environmental interpretation being carried out on each well so as to locate the next one in optimum position. Nowadays however seismic surveys make it possible to map the geometry of facies long before any well has been drilled. Space does not allow the seismic method to be outlined here, but this topic is covered in many text books (e.g. Coffeen, 1978, Kleyn, 1982 and Anstey, 1982). A whole new branch of seismic interpretation has now evolved known loosely as seismic stratigraphy. This applies established

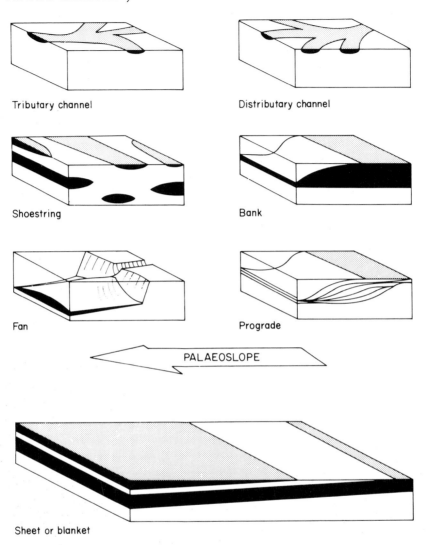

*Figure 1.3* Cartoons of various facies shapes. The diagnostic significance of the various geometries are discussed in the text.

stratigraphic and facies concepts to the interpretation of seismic data. Seminal papers on this topic by Vail, Mitchum and their colleagues will be found in an important volume edited by Payton (1977).

Examination of a seismic line often shows that it may be divided into a number of units. The geometry of these units may be mappable using the data from a complete seismic survey. Channels, fans, progrades and many other forms may be detectable. The shapes of these seismic units are often environmentally diagnostic as previously discussed. The character of seismic unit boundaries may be important in establishing the geological history of the area. Vail, Mitchum *et al.* (1977) have defined the various types. Seismic unit boundaries may be either concordant or discordant. The former are rare because where reflectors are concordant it will be difficult to map a boundary unless there are significant differences in the internal seismic character of adjacent units.

Discordant unit boundaries are of several types. The upper boundary will show offlap or, as it is sometimes termed, toplap. Seismic data alone may not show whether this boundary is due to erosional truncation, or whether it reflects a regressing increment deposited while sea level remained constant. The lower discordant boundary of a seismic unit may show either onlap or downlap. Onlap is commonly caused by a marine transgression, or by the progressive infilling of a continental basin. Downlap is generally caused by progradion, either of a whole shoreline or of an individual fan of some sort. Fig. 1.4 illustrates the various types of seismic unit boundary.

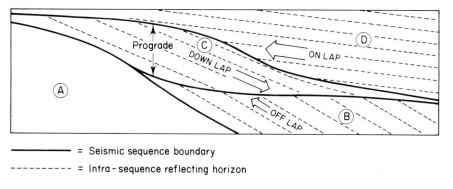

——————— = Seismic sequence boundary

-  -  -  -  -  -  -  - = Intra - sequence reflecting horizon

*Figure 1.4* Sketch to show some of the terminology of seismic stratigraphy.

The internal nature of a seismic unit may be as instructive as its' boundaries. The two important parameters to consider are the amplitude and continuity of the reflecting horizons (Anstey, 1982, p. 26–30). High amplitude reflectors with good lateral continuity commonly indicate interbedded marine limestones and shales. High amplitude events of poor continuity characterize continental deposits with interbedded channel sands and shales. Low amplitude reflectors with good continuity suggest basinal shale sediments. Hummocky reflectors may be produced from the corrugated topography of shallow marine and coastal sand bodies. Chaotically dipping reflectors of short continuity may indicate slumped beds, especially if seen at the base of a prograde.

## Lithology

The lithology of a sedimentary facies is one of the easiest of parameters to observe and one of considerable environmental significance. As a general rule the lithology of limestones is of more diagnostic importance than that of sandstones. This is because studies of Recent carbonates show that they occur in a number of micro-facies whose distribution is closely related to their depositional environment (e.g. Imbrie and Purdy, 1962). Furthermore, they cannot survive transportation far from where they formed (Ginsburg and others, 1963, p. 572). Because of these facts many limestones can be attributed to an environment by examination of a small chip or thin section and close comparison with Recent deposits (see Chapters 8 and 9).

The lithology of a clastic sediment on the other hand is a function not only of the environment in which it was deposited but also of its transportational history and of the type of rock from which it was derived. Petrographic studies of sandstones are thus of much less value as an indicator of depositional environment than those of carbonates. On the other hand sandstones are less susceptible to diagenesis than carbonates, so their depositional fabric is generally easier to discern.

Many attempts have been made to use the texture of a sediment to determine its depositional environment. This topic is described in a vast literature which can scarcely be avoided by leafing through any recent textbook or journal dealing with sedimentary rocks. Folk (1967), Moiola and Weiser (1968), Pettijohn *et al.* (1972, pp. 86–8) and Selley (1982a, pp. 19–22) have reviewed the problems of this approach. The basic philosophy of this technique is that the sediments of Recent environments can be distinguished from one another by careful statistical evaluation of their texture. Numerous studies have shown just how easily this may be done, e.g. Feldhausen and Ali (1975), and Reed *et al.* (1975). By analogy it should be possible to carry out granulometric analyses of ancient sediments, subject them to various statistical gymnastics and, by comparing them with Recent sediments of known origin, determine their depositional environment.

Unfortunately this approach has largely proved unsatisfactory and the granulometric analysis of ancient sediments is a declining art, to the delight of laboratory technicians. There are several reasons for the failure of this method. First, as already mentioned, the texture of a sediment is a function not just of the depositional environment, but also of its previous history. Thus, to take an extreme case, if a clean well-sorted fine sand is the only sediment being transported into an environment then a clean well-sorted fine sand is all that can be deposited in it. A second problem of using texure to determine the environment of a sediment lies in the origin of the fine detritus. Studies of Recent sediments show that clay content is a very sensitive indicator of depositional process (Doeglas, 1946). In an ancient sediment it is not possible to prove that the clay matrix was deposited at the same time as the rest of the detritus. It is also possible that it was washed in at a later date, that it was transported in particles coarser than clay, or that it was formed from the diagenetic breakdown of chemically unstable sand grains (Cummings, 1962). A third problem of the granulometric analysis of sedimentary rocks is a purely technical one.

Post-depositional solution and overgrowths may considerably alter the size of sand grains and modify the over-all texture of a rock. Intensive cementation may make it impossible for a rock to be disaggregated to its original textural composition. In such cases it may be necessary to determine the sorting curve by measuring the dimensions of grains in thin sections. This is a slow and complex process and the end results are extremely difficult to compare with sieve analyses of Recent sediment samples (Chayes, 1956).

For the reasons outlined above, therefore, statistical textural studies of ancient sediments have largely proved an unsatisfactory method of environmental diagnosis, though some success stories have been recorded (Visher, 1969).

Considered from a more general standpoint the grain size of a sediment is an important indicator of the energy level of a depositional environment. The coarser the grain size the higher the energy level of the depositing current and the better the sorting the more prolonged its action. These generalizations have long been applied to terrigenous sediments and form the basis for a useful classification of carbonates (Dunham, 1962). However even these generalizations of the correlation of grain size and sorting to energy level must be interpreted with caution. No matter how strong a current it cannot deposit sediment coarser than that available in the source. Similarly the grain size of carbonate rocks may not always be a reliable indicator of the energy level of their depositional environment. This is because many limestones are composed of skeletal particles. Shells can form what are virtually organic conglomerates in a low-energy environment merely by living and dying there. Lagoonal oyster reefs are a case in point. Similarly the sorting and micrite content of carbonates is not solely a function of turbulence but is also influenced by other factors, notably micrite-forming algae and shell-munching predators (Bathurst, 1967, pp. 463–4).

Considerable attention has been given to the way in which depositional processes may affect the shape of sand grains. Recent glacial sands tend to be more angular and of lower sphericity than water-borne grains, while dune sands are often very well rounded. Kuenen (1960) summarizes a series of papers describing the experimental abrasion of sand grains by various processes. His data confirm that wind action is a more effective rounding agent than running water. Grain shape however is not just a function of the last depositional process which affected it but also of its previous history and original shape. Polycyclic sands will tend to be well-rounded regardless of the various processes to which they have been subjected (see also p. 92).

Electron microscope examination has shown that glacial, aqueous and eolian processes imprint characteristic markings on grain surfaces (e.g. Krinsley and Funnel, 1965, Krinsley and Cavallero, 1970, and Krinsley and Doornkamp, 1973). This is a valuable technique but it must be interpreted with care. Post-depositional solution, compaction, and tectonic grinding may modify the surface texture of a sand grain. It may also take a while for a grain which has been transported from one environment to another to have its surfaces altered.

Nonetheless eolian transport textures have been observed on sandstones as old as Triassic (Krinsley *et al.*, 1976).

Apart from its physical aspects the chemistry of a sediment holds many important clues as to its origin. Degens (1965) gives a comprehensive account of this topic. Only one or two examples will be described here to illustrate the present scope and future potential of geochemistry in environmental diagnosis.

There has been considerable discussion on the use of clay minerals as indicators of sedimentary environments (e.g. Grim, 1958; Shaw, 1980). As a general rule, however, the chemistry of a clay reflects not just its depositional environment but also the parent rock, the climate which weathered it, and its diagenetic history. Weaver (1958) on the basis of hundreds of analyses of clays from ancient sediments showed that no single clay mineral is specific to any one environment, even in such broad terms as marine and non-marine. His data do show however that illite and montmorillonite are more characteristic of ancient marine rocks, while kaolinite is more typical of continental, especially fluviatile deposits (ibid., p. 258, Fig. 1).

Glauconite is a mineral which is widely believed to form only in marine environments and to be so unstable that it cannot survive re-working. It is therefore held to be diagnostic of marine rocks. This criterion is not infallible. Detrital glauconite occurs in fluviatile Neogene red beds in the Dead Sea Valley. As a general rule however glauconite is a useful indicator of marine environments and studies of the present day distribution of glauconite suggest that it may be possible to narrow down the precise conditions which limit its formation (Ireland *et al.*, 1983).

The Boron content of clays was first discussed as a palaeosalinity indicator by Degens and others in 1957, and has subsequently been studied by many workers. Analysis of 82 Recent muds has shown that, for a given clay content, marine muds contain 30–45 ppm more Boron than freshwater muds (Shimp *et al.*, 1969). This technique for distinguishing non-marine and marine deposits has been applied to ancient rocks with varying degrees of success (see for example Potter *et al.*, 1963, and Murty, 1975).

Similar geochemical approaches have been applied to other constituents of Recent sediments, notably phosphates and iron, in attempts to diagnose their depositional environments in ancient rocks. These studies have not so far been notably successful due largely to the scarcity of Recent environments where phosphates and iron are forming (see for example Bromley, 1967 and Rohrlich *et al.*, 1969, respectively).

In conclusion it can be seen that the lithology of a sedimentary facies holds many important clues to its depositional environment. This is truer of carbonates than sandstones since they are deposited at, or close by, their point of origin. The lithology of sandstones gives less indication of their depositional environment since the sediment is introduced from outside the site of deposition and inherits extraneous characteristics due to its previous history.

Grain size, sorting, shape, and texture often reflect the energy level and process of the environment. They must be interpreted carefully. Geochemistry, a rapidly expanding field, has great potential in environmental diagnosis.

## Sedimentary structures

Sedimentary structures are very important indicators of depositional environment. Unlike lithology and fossils they are undoubtedly generated in place and can never have been brought in from outside. Sedimentary structures are easy to study at outcrop where exposure is good. In sub-surface work, however, only the smallest ones can occasionally be found in rock cores.

A huge number of sedimentary structures have been described in the literature. Their nomenclature and classification is confused and complex. This is largely because it is extremely hard to define accurately their morphology. Atlases of sedimentary structures have, however, been compiled by Pettijohn and Potter (1964), Gubler and others (1966), and Conybeare and Crook (1968).

Within recent years the interpretation of sedimentary structures has been greatly enhanced both by experimental work and by studies of Recent environments (Allen, 1970; Raudkivi, 1976; Selley, 1982a, pp. 169–95; and White *et al.*, 1976).

Sedimentary structures can provide evidence of whether an environment was glacial, aqueous, or sub-aerial. They give some indication of the depth and energy level of the environment and of the velocity, hydraulics, and direction of the currents which flowed across it (Allen, 1967, 1982; Collinson and Thompson, 1982; Reineck and Singh, 1980; and Selley, 1982, pp. 208–54).

Most sedimentary structures can be arbitrarily fitted into a genetic classification of pre-, syn- and post-depositional categories. The environmental significance of sedimentary structures will now be briefly reviewed under these three headings.

### *Pre-depositional sedimentary structures*

Pre-depositional sedimentary structures are those observed on bed interfaces which formed before deposition of the younger bed. They are thus erosional features, not to be confused with post-depositional bed base deformational phenomena such as loadcasts. Pre-depositional structures include channels, scour marks, flutes, grooves, tool markings, and a host of other erosional phenomena. Despite considerable experimental work the hydraulic conditions which cause these structures are not well understood. Regardless of this they give valuable indications of the flow directions of the current which generated them.

### *Syn-depositional sedimentary structures*

Structures formed during deposition include flat-bedding, crossbedding, lamination, and micro-crosslamination (ripple-marking).

When a bed of sand is subjected to a current of increasing velocity different bed forms develop in a regular sequence. Each bed form deposits a different sedimentary structure (see Table 1.3). The point at which one bed form changes

*Table 1.3. The sequence of bed forms and sedimentary structures produced by a current of increasing velocity passing over a sand bed. The threshold velocity at which one bed form changes to another increases with increasing grain size. The ripple phase is absent for sands coarser than about 0·60 md. (Based on data in Simons, Richardson, and Nordin, 1965.)*

| | | Flow regime | Bed form | Sedimentary structures |
|---|---|---|---|---|
| ↑ | Erosion | | | |
| | | Upper | Antidunes (upcurrent migrating dunes) | Seldom preserved due to re-working as velocity decreases |
| Increasing current velocity | Deposition | Transition | Plane bed | Flat-bedding |
| | | Lower | { Dunes { Ripples | Cross-bedding Micro-crosslamination |

Non-deposition: Current velocity too low to transport sediment.

to another is a function of many variables, including the velocity, temperature, and viscosity of the fluid and the fall velocity (which more or less corresponds to the grain size) of the sediment particles. These variables have been integrated into the flow regime concept which expresses the sum total of all these parameters (Simons, Richardson, and Nordin, 1965). The main value of the flow regime concept is that it points out that beds of the same grain size need not necessarily have been deposited by currents of equal velocity. For example a micro-crosslaminated sand was deposited by a current of lower velocity than that which deposited a flat-bedded sand of equivalent grade. Similarly two beds of different grain size may not have been deposited by currents of different velocities. A flat-bedded fine sand and a cross-bedded medium sand may have been deposited by currents of the same velocity.

Despite the valuable insight that the flow regime concept has given to the study of syn-depositional sedimentary structures it has only indirectly aided environmental analysis. This is because the conditions generated by the confined flow of water in a sediment flume occur in nature in many different situations, such as rivers, estuaries, delta distributaries, tidal channels, and so on.

Cross-bedding is a sedimentary structure which is both common, morphologically variable, and extensively documented (see Potter and Pettijohn, 1963, pp. 62–89). Much can be learnt about the process which formed cross-bedding, but, since few processes are restricted to any one environment, this knowledge is of limited value (Allen, 1964, 1982). Large scale eolian cross-bedding may be an exception to this rule (see p. 95).

Micro-crosslamination however is of more use as an environmental indicator. Walker (1963) and Jopling and Walker (1968) have described morphological criteria for distinguishing ripples due to different processes which are sometimes environmentally specific. Tanner (1967) has empirically determined a number of statistical criteria for identifying the depositional environment of ripples. Even so, Allen (1968, 1982) has shown how the most sophisticated of analytical and experimental techniques, while contributing much to our understanding of ripple formation, are of little value in environmental interpretation.

*Post-depositional sedimentary structures*
Structures which develop in sediments after their deposition are no less diverse and complex than those which form during deposition. They too have generated a ponderous literature in geological books and journals (see Mills, 1983 for a useful review). Again experimental studies, and the examination of Recent deposits have done much to increase our understanding of soft sediment deformation. Basically it is possible to distinguish two main genetic groups of these structures. One group shows an essentially vertical reorientation of bedding, while the second shows lateral rearrangement of the fabric. The first type includes loadcasts and pseudonodules where beds of sand collapse into underlying mud, as well as convolute lamination and quicksand structures which are contortions within beds of interlaminated fine muddy sand and clean sand respectively. Experimental studies have shown that there are a variety of triggering mechanisms which can generate these structures in unconsolidated sediment. These include earthquakes, current turbulences, and hydrostatic pressures due to connate fluids (see Selley, 1969a for a review). These processes can all operate in a variety of geological situations and are therefore of little significance in environmental analysis.

The second group of post-depositional sedimentary structures are those due to lateral movement of sediment, as opposed to the predominantly vertical reorientation shown by the previous type. These structures include slumps and slides which are penecontemporaneous recumbent folds and faults indicating large-scale sideways transport of sediment. Such phenomena occur where rapid sedimentation or erosion forms steep slopes which, from time to time, become so unstable that they collapse and shed superficial deposits down slope. These slope failures can occur spontaneously or be triggered by earthquakes, storms, or herds of stampeding dinosaurs.

Deformational sedimentary structures due to lateral movement can occur in many geological situations ranging from river banks to abyssal slopes. Slumping on an extensive scale however seems to be most typical of delta slopes and submarine canyons and fans.

Regardless of whether slumps can be used to determine depositional environment they are important criteria for recognizing palaeoslopes.

## Palaeocurrent patterns
Of the five defining parameters of a sedimentary facies (geometry, lithology, structures, palaeocurrents, and fossils) all but palaeocurrents are observational criteria. To determine the palaeocurrents of a facies involves not just the description, but also the interpretation of data. This is a severe handicap.

Compensation is to be found however in the fact that since palaeocurrents are discovered from sedimentary structures they clearly reflect the depositional environment of a facies and cannot inherit features from outside the actual site of deposition. The palaeocurrent analysis of a facies involves the following steps:

(*i*) Measurement of the orientation of significant sedimentary structures in the field (e.g. cross-bedding dip direction, channel axes, etc.).

(*ii*) Deduction of palaeocurrent direction at each sample point.

(*iii*) Preparation of regional palaeocurrent map.

(*iv*) Integration of palaeocurrent map with other lines of facies analysis to determine environment and palaeogeography. In some environments palaeocurrents may indicate palaeoslope (e.g. in rivers), in others they do not (e.g. eolian deposits).

The methodology and applications of palaeocurrent analysis have been reviewed by Potter and Pettijohn (1977) and Selley (1982a, pp. 231–40). Many workers have warned that the deduction of palaeocurrent directions must be carried out with great care, both because of the complex relationship between structures and currents, and the diverse and variable nature of the currents themselves (e.g. Allen, 1966). None the less it has been pointed out that for Recent sedimentary environments there is a close correlation between their current systems and the local and regional orientation of their sedimentary structures (Klein, 1967). Similarly Selley (1968) has suggested that there appear to be a number of palaeocurrent patterns which have been recorded repeatedly from rocks of different ages all over the world. These different patterns or models are specific to various environments. Each can be defined both by the

Table 1.4. *A classification of some palaeocurrent patterns. Based on data in Selley (1968).*

| Environment | | Local current vector | Regional pattern |
|---|---|---|---|
| Alluvial | Braided | Unimodal, low variability | Often fan-shaped |
| | Meandering | Unimodal, high variability | Slope-controlled, often centripetal basin fill |
| Eolian | | Uni-, bi- or polymodal | May swing round over hundreds of miles around high pressure systems |
| Deltaic | | Unimodal | Regionally radiating |
| Shorelines and Shelves | | Bimodal (due to tidal currents), sometimes unipolar or polymodal | Generally consistently oriented onshore offshore, or longshore |
| Marine turbidite | | Unimodal (some exceptions noted, see p. 253) | Fan-shaped or, on a larger scale, trending into or along trough axes |

palaeocurrent vector at individual sample stations and by their regional relationships (Table 1.4).

It would seem therefore that, though it must be interpreted with great care, palaeocurrent analysis is an important technique for recognizing ancient sedimentary environments and their palaeogeographies.

## Fossils

Last, but certainly not least, of the five parameters defining a facies are its fossils. These have always been one of the most important methods of identifying the depositional environment of a sediment. The way in which fossils lived, behaved towards one another, and influenced and were influenced by their environment is termed palaeoecology. Reviews of this large field of research have been given by Hedgpeth and Ladd (1957), Gecker (1957), Ager (1963), Imbrie and Newell (1964), Dodd and Stanton (1981), McCall and Tevesz (1982) and Gall (1983). To use fossils to identify the depositional environment of the host sediment two assumptions must be made:

(1) That the fossil lived in the place where it was buried.

(2) That the habitat of the fossil can be deduced either from its morphology or from studying its living descendants (if there are any).

These are two very real problems which must always be kept in mind when using fossils as environmental indicators. It is not always easy to be sure that a creature lived in or on the sediment in which it was buried. Many fossils are preserved in a particular environment not because they lived in it but because they found their way into it by accident and it was so hostile that it killed them. Think of all the drowned cats washed out to sea by the River Thames.

The second problem, that of deducing the habitat of a fossil, is also a very real one which has been discussed extensively in the literature. To consider one simple case, it has been pointed out that bears live today from the arctic to the equator. If only Polar bears lived at the present time old bear bones might be used as indicators of glacial climates (Shepard, 1964, p. 4).

Lest the preceding paragraphs have painted too gloomy a picture of fossils as environmental indicators it must be stated that they are one of the most valuable techniques available.

Of all the different fossils that can be used in environmental analysis perhaps two of the most important types are micro-fossils and trace fossils. Micro-fossils have the great advantage over mega-fossils that they are recoverable from well cuttings, and that a small volume of rock may contain sufficient specimens to be used in statistical studies. There are many different groups of microfossils that may be used in environmental interpretation. These include the foraminifera, ostracods, microplankton and palynomorphs. One of the problems faced in environmental interpretation is that different groups of microfossils may give conflicting answers. The average geologist generally lacks the specialist expertise necessary to adjudicate in such disputes. One of the most useful groups of fossils for environmental diagnosis is burrows, tracks, and trails, collectively called trace fossils. These have been extensively studied in Crimes and Harper (1970, 1977) and Frey (1975).

Trace fossils are useful in environmental interpretation for two reasons. First, they occur *in situ*, and cannot therefore be reworked like other fossils. Second, it has been noted that certain types of trace characterize particular environments, though the beasts responsible may be unknown. Using this fact workers have defined a series of 'ichnofacies'. Each ichnofacies consists of a suite of trace fossils which occur in characteristic sedimentary facies whose environment may be determined independently (see Seilacher, 1967; Rodriguez and Gutschick, 1970; and Heckel, 1972). These ichnofacies have remained constant through Phanerozoic time, despite the evolution of the creatures responsible (Fig. 1.5).

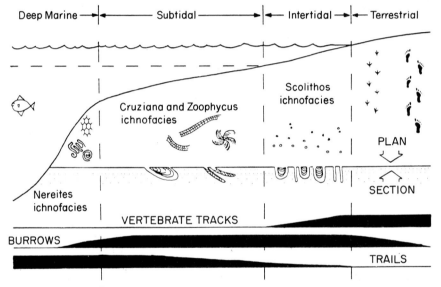

*Figure 1.5*  Characteristic ichnofacies for various environments. From Selley (1976, Fig. 7.1) by courtesy of Academic Press Inc.

For this reason, and because of their preservation *in situ*, trace fossils are very useful in sub-surface facies analysis when they are found in cores.

Used carefully these and other fossil groups indicate many environmental parameters including depth, temperature, salinity, current turbulence, and climate. This necessarily brief résumé of a complex problem is discussed in greater depth in the references quoted at the beginning of this section.

## SUB-SURFACE ENVIRONMENTAL INTERPRETATION

The preceding sections of this chapter have reviewed how depositional environment may be deduced from rocks at the surface using the five facies-defining parameters (geometry, lithology, sedimentary structures, palaeontology, and palaeocurrents). The chapter concludes with a review of the methods which may be used to interpret the depositional environment of sediments in the sub-surface. The concepts are unaltered for this purpose but the techniques are substantially different.

Considering geometry first, it has already been discussed that whereas, for example, reefs and channels may be recognized from their exhumed topography at outcrop, in the sub-surface the geometry of a facies is identified from borehole data or seismic maps.

Lithology is of varied utility in sub-surface environmental interpretation, and the difference of the methods applied to carbonate and terrigenous rocks cannot be overemphasized. As already mentioned, and as will be elaborated in Chapters 7, 8, and 9, there are various types of carbonate grain which form in different environments. Therefore the origin of carbonate rocks in the sub-surface is largely deduced from petrographic studies of well samples. The environment of deposition of terrigenous rocks may seldom be deduced from petrography, however, because of the problems of inheritance and diagenesis discussed earlier. The key to the sub-surface diagnosis of terrigenous rocks lies in the study of vertical sequences: sequences of sedimentary structures in cores, and sequences of grain size which may be inferred from geophysical logs.

After a borehole has been drilled the geophysical properties of the adjacent strata are measured by passing a series of electronic devices along the borehole. These can record many of the physical and chemical properties of a sediment, including resistivity, density, sonic velocity, and radioactivity. These logs can be used to correlate adjacent wells, to identify lithology, to measure the amount of porosity, and to calculate the amount of oil, gas, or water within the pores.

The study of formation evaluation, though an essential tool for stratigraphers and sedimentologists, is beyond the scope of this book and the reader is referred to works by Asquith (1982) and Pirson (1983) for further details. Only two aspects of formation evaluation are essential in the present context; the use of geophysical logs as vertical profiles of grain size, and the use of the dipmeter log in the interpretation of sedimentary dips and palaeocurrent analysis.

## Interpretation of vertical grain-size profiles from geophysical logs

When discussing cyclic sedimentation earlier in this chapter it was pointed out how depositional systems often tend to deposit characteristic sequences of grain size and sedimentary structures.

The response of several geophysical logs is indirectly related to sediment grain size. Such logs may thus be used as vertical grain-size profiles. This is a well established technique which has been widely used in the oil industry for many years (see for example Galloway, 1968; Fisher, 1969; Pirson, 1983; and Selley, 1976b). The two logs which are mainly used as grain-size profiles are the Spontaneous Potential (generally abbreviated to S.P.) and Gamma logs.

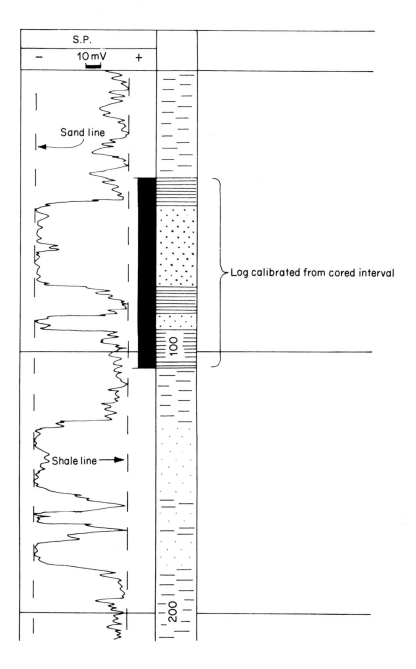

*Figure 1.6* Well log of spontaneous potential (S.P.) curve against depth, showing how sand line and shale base lines are constructed. Sand grain size can be calibrated with the S.P. curve where cores or well cuttings are available.

The S.P. log records the difference between the potential of an electrode moved along the borehole and the fixed potential of an electrode at the surface. This varies with the cumulative effect of electrofiltration and electro-osmosis in the strata adjacent to the borehole. These parameters are essentially related to permeability. It is a general observation that shales are impermeable and that sands are permeable. Furthermore in sands permeability tends to decrease with grain size (see for example Pryor, 1973). This is because the amount of clay matrix tends to increase with decreasing grain size, and the matrix blocks the throat passages between pores. Thus because the S.P. log essentially records permeability and because permeability is generally related to particle size, then this log may be used as a continuous vertical grain-size profile.

If a well has been cored then a direct calibration can be made between observed grain size in the core and the adjacent S.P. value (measured in millivolts). Even when there is no cored interval it is generally possible to establish a 'sand line' and a 'shale base line' from the two extreme values of the S.P. curve (Fig. 1.6). Calibration may then be made since the sand line will correspond to the coarsest sand samples seen in the well samples.

Though the use of the S.P. log as a grain-size profile is a well-established technique it has several limitations naturally. The relationship between the permeability and grain size is valid only for sediments with primary intergranular porosity.

Thus the S.P. log may not reflect grain size in cemented sandstones or in carbonates with complex diagenetic histories. Furthermore the S.P. curve only shows good amplitude when there is a substantial difference between the salinity of the drilling mud and that of the formation waters. For a number of reasons S.P. curves of offshore wells seldom show sufficient amplitude to be used as grain-size curves.

The second geophysical tool which may be used for this purpose is the gamma log. This records the radioactivity of the formation as the sonde is passed along the borehole. This is essentially related to the amount of clay present, and this largely occurs in the clay minerals. Thus the gamma reading tends to increase in proportion to the amount of clay present in the sediment, and as already stated, this tends to increase with decreasing particle size. Thus the gamma log, like the S.P., can be used as a grain-size profile. Sand line and shale base line values can be calibrated to particle size in the same way.

Like the S.P. log, the gamma tool has several limitations. It is affected by hole diameter and may give too low a reading where there is extensive caving. This can be monitored by the caliper log which is generally run simultaneously with the gamma, and a correction can be applied. A second problem is the occurrence of radioactive minerals other than those present in clays. In particular the presence of large amounts of mica, glauconite, or zircon can make a sand abnormally radioactive and appear fine and shaley on the gamma log.

There is one particular trick that the gamma log can play. This concerns channels with basal conglomerates. With an extraformational conglomerate there

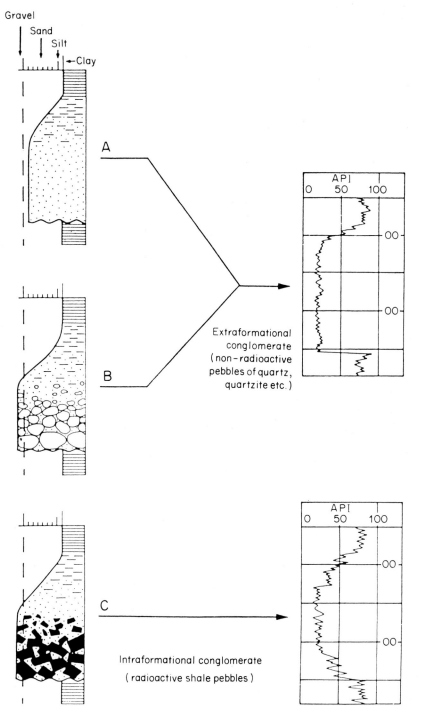

*Figure 1.7* Grainsize profiles and gamma log responses for some upward-fining sequences. Note how intraformational conglomerate on a channel floor may cause the gamma log to apparently indicate an upward-coarsening sequence.

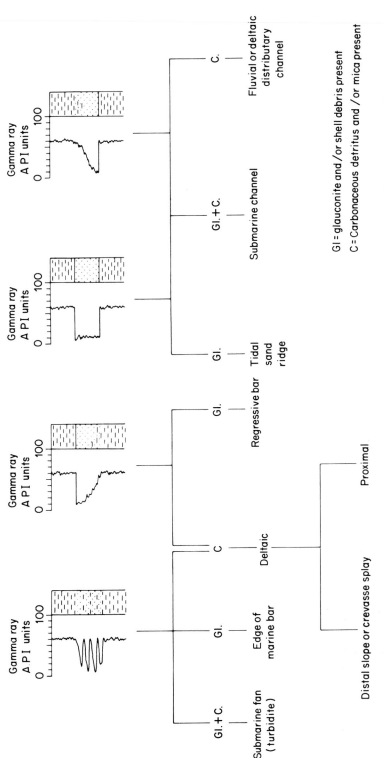

*Figure 1.8* Four characteristic gamma log motifs. From left to right: thinly interbedded sand and shale; an upward coarsening profile with an abrupt upper sand : shale contact; a uniform sand with abrupt upper and lower contacts; and, furthest right, an upward fining sand : shale sequence with an abrupt base. None of these motifs is environmentally diagnostic on its own. Coupled with data on their glauconite and carbonaceous detritus content, however, they define the origin of many sand bodies.

is generally no problem. Pebbles of quartz, quartzite, etc. give the same low API readings as terrigenous sand. Where a channel is floored by an intraformational shale pebble conglomerate the situation is very different. The gamma log may apparently indicate an upward-coarsening grain size motif when in fact the very opposite is the case (Fig. 1.7).

The gamma log has a limited use in carbonate facies analysis where it can indicate fluctuations in clay content, but it does not vary with carbonate grain size *per se*.

Despite their limitations both the S.P. and gamma logs can often be used as vertical grain-size profiles. This technique is very useful sub-surface analysis as is shown in this book.

It cannot be too strongly stressed, however, that the starting point for all sub-surface facies analysis is the meticulous sedimentological logging of every available core. Thus the wireline log can be calibrated with real rocks and they can be related to the associated lithologies, textures and structures.

Figure 1.8 shows some of the characteristic gamma log motifs which can commonly be seen. These, from left to right, indicate: thinly interbedded sand and shale, a sand body with a gradational base and abrupt top (i.e. with an upward-coarsening grain-size profile), a uniform sand body with abrupt upper and lower contacts, and lastly, a sand with an abrupt base and a gradational top (i.e. with an upward-fining grain-size profile).

While it is true that clastic environments deposit characteristic grain-size profiles, yet no environment possesses a unique motif. Similar motifs are produced by different environments. For this reason wireline log patterns must never be interpreted in isolation. As already pointed out, accurate diagnosis can more easily be made where supplementary core and palaeontological data are available. Where these are absent or inconclusive the occurrence of glauconite, shell debris, mica and carbonaceous detritus can be used to assist in the correct identification of log motifs.

Glauconite is a complex mineral related to the clays and the micas; essentially it is an aluminosilicate containing magnesium, iron, and potassium. Glauconite occurs as a replacement of faecal pellets and as an internal mould of forams and other small voids. Detailed discussion of the genesis of glauconite in modern sediments will be found in Odin and Matter (1981), Logvinenko (1982) and Ireland *et al.* (1983).

The consensus of opinion is that glauconite forms only as an authigenic mineral during the very early diagenesis of marine sediment. Penecontemporaneous reworking can concentrate glauconite in shoal sands and transport it into deeper basinal sands.

Glauconite is extremely susceptible to sub-aerial weathering and, except in one or two extremely rare instances, it is not known as a reworked second cycle detrital mineral. For this reason the presence of glauconite in a sand is a useful criterion for a marine origin.

The absence of glauconite is, of course, inconclusive and does not indicate a non-marine environment. It is interesting to note that on a world wide basis that glauconite formation was very common in the Lower Palaeozoic and from the Jurassic through to the Palaeocene (Pettijohn, Potter, and Siever, 1972, p. 229).

The great attraction of glauconite is therefore that it can be used as an indicator of a marine environment. It can be identified from cuttings in sandstones where there are no cores and which are palaeontologically barren. Correct identification is of course imperative, and glauconite should be distinguished from green, but flaky, chlorite and from waxy green weathered volcanic grains.

Identifiable fossil fragments are obviously very important environmental indicators. It is possible however, that the mere presence of unidentifiable shell fragments in a sandstone may suggest a marine environment. Certainly lime-secreting organisms occur in both marine and non-marine environments. It appears however that the preservation potential of their skeletal remains is lower in continental sediments than marine ones. This may be because in the former calcareous shell fragments are likely to be leached out by percolating acidic ground water. In marine sediments however early diagenesis tends to occur in a neutral or alkaline pore fluid. Thus lime may be preserved, and even precipitated. Most geologists would probably support the statement that shelly sands tend to be marine rather than continental.

The third important authigenic constituent of sands which can help define environment is carbonaceous detritus. Microscopic particles of lignite and coal are a common minor component of many sands. Traditionally this is taken as a criterion of a non-marine or deltaic environment. This is not invariably true. It is unlikely that many of the coals found in carbonate sequences are terrestrial in origin. More probably they are due to marine algae. Similarly masses of rotting seaweed are reported from modern submarine canyons down which they are transported, mixed with shelf derived glauconite, to be deposited in submarine fans.

The presence or absence of carbonaceous detritus is not therefore an indicator of non-marine or marine environments. It is a winnowing index which reflects the degrees of turbulence and agitation to which a sediment has been subjected.

Detrital mica flakes can be used in much the same way as carbonaceous detritus. Mica is generally winnowed out of high energy environments and tends to be deposited in low energy ones where sediment is rapidly deposited with little reworking. Thus micaceous sands characteristically occur in crevasse-splays, outer delta slopes and in the upper parts of turbidites.

Figure 1.9 shows how the presence or absence of glauconite shell debris, mica and carbonaceous detritus define four major groups of sands.

Glauconitic and/or shelly sands devoid of carbonaceous detritus are formed in the high energy marine environments of sand bars, shoals, and barrier islands. Carbonaceous and/or micaceous sands, from which glauconite and shell debris are absent, characterize deltaic, fluvial, and lacustrine environments. The mix

| | | Marine | Non-marine |
|---|---|---|---|
| | | Glauconite and/or shell debris | No glauconite or shell debris |
| Well winnowed | Not carbonaceous detritus or mica | barrier<br><br>bar<br><br>shoal     }   marine shelf sands | eolian |
| Poorly winnowed | Cabonaceous and/or micaceous | submarine channel<br><br>and fan | fluvial<br><br>lacustrine<br><br>deltaic |

*Figure 1.9*   The presence or absence of glauconite and carbonaceous detritus divides sands into four main environmental groups: marine winnowed sands, non-marine winnowed sands (eolian), mixed sands with both glauconite and carbonaceous matter, commonly found as turbidites, and poorly winnowed non-marine sands.

of all four constituents is especially characteristic of deep sea environments in which shelf-derived shelly glauconite sand is dumped with mica flakes and humic or algal organic debris. In an 'ideal' situation a sand with none of these constituents would be eolian, but this diagnosis must be treated with scepticism unless supported by adequate supplementary data.

This technique is real cowboy geology which must never be used in isolation. It is extremely interesting however to combine this approach with the analysis of wireline log motifs.

Figure 1.8 shows how the log motifs, non-diagnostic on their own, can be allocated to specific environments when data are available on the presence or absence of glauconite and carbonaceous detritus.

Consider first the nervous 'ratty' log motif, indicating thinly interbedded sands and shales. If these sands contain both glauconite and carbonaceous detritus then, as already argued, this is probably a submarine fan or turbidite deposit. The presence of glauconite and/or shell fragments on their own may indicate that the sand was laid down on the flank of a bar sand.

The presence of mica and/or carbonaceous detritus on their own suggests an origin in a distal deltaic or crevasse-splay environment.

Consider now the second log motif, reflecting an upward-coarsening grain-size profile. If such a sand is carbonaceous a seaward prograding delta lobe is the preferred environment. If on the other hand this is glauconite, then it is probable that this is a regressive sand bar.

The third log motif, reflecting a sand with abrupt upper and lower contacts and a uniform grain-size profile, is open to three interpretations. If it contains glauconite and/or shell debris alone, then it is most probably a marine shoal sand or tidal sand ridge. These sands have erosional bases, abruptly overlying an abraded substrate. They are well sorted, contain skeletal detritus and lack clay and carbonaceous matter.

A sand of uniform grain size with abrupt upper and lower contacts, containing mica and/or carbonaceous matter and no glauconite or shell debris, is probably a channel, either a fluvial channel or a delta distributary. Such a sand body with both glauconite and/or shell fragments together with mica and/or carbonaceous matter is probably a submarine channel which acted as a feeder for a submarine fan at its basinward mouth.

Consider now the last characteristic log motif: a sand with an abrupt base and a transitional upper contact. Upward-fining grain-size profiles such as this characteristically reflect the gradual infilling of a channel.

The presence of mica and/or carbonaceous detritus on their own suggest a fluvial or deltaic environment. Where these are present together with shell debris and glauconite this is probably a submarine channel.

Thus the study of log motifs combined with the study of petrography can be very useful in the detection of sand body environments.

## The use of the dipmeter in sub-surface facies analysis

Another important technique in sub-surface facies analysis is the interpretation of sedimentary dips from the dipmeter log. The dipmeter tool consists of a sonde with four pairs of closely spaced electrodes at 90° from one another. As the sonde is withdrawn from the borehole the resistivity of the formation is measured at the four opposed points on the borehole wall, and is recorded electronically on four tracks. Simultaneously a recording is made of the orientation of the sonde with respect to the magnetic north.

The resistivity of sediments is very variable, so the four traces show a very erratic character. By selecting four similar events on the curves, which relate to a bedding surface, it is possible to calculate the direction and the amount of dip of the surface. This is done by computer, which selects events, calculates the azimuth and amount of dip, and gives a statistical appraisal of the significance of each reading (Fig. 1.10).

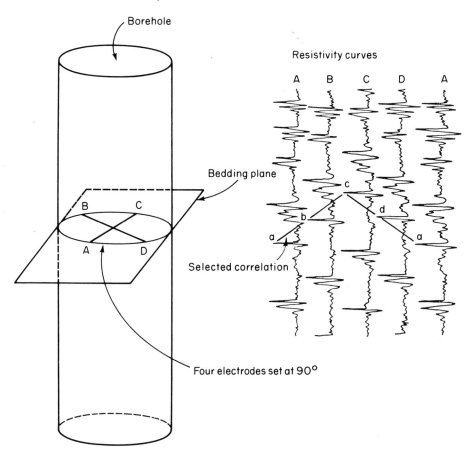

*Figure 1.10*   Sketch to show the mode of operation of the dipmeter log and the way in which dip directions are calculated from the four mutually opposed resistivity curves.

Dip results may be presented in tabular form, but are more commonly shown by one of several graphic displays. That which is most frequently used is the 'tadpole' plot, in which each reading is plotted as a dot on a graph whose vertical ordinate is depth and whose horizontal ordinate shows the amount of dip. A short line appended to each dot points in the direction of dip (Fig. 1.11).

More detailed accounts of the operation, computation, and display of dipmeter logs will be found in the manuals of well logging service companies.

The dipmeter tool is invaluable in deciphering sub-surface structures in areas of complex tectonics. Folds, faults, and unconformities may be identified and their orientation determined from the dipmeter. Such applications of the tool are beyond the scope of this book.

The dipmeter is extremely relevant however to sub-surface facies analysis.

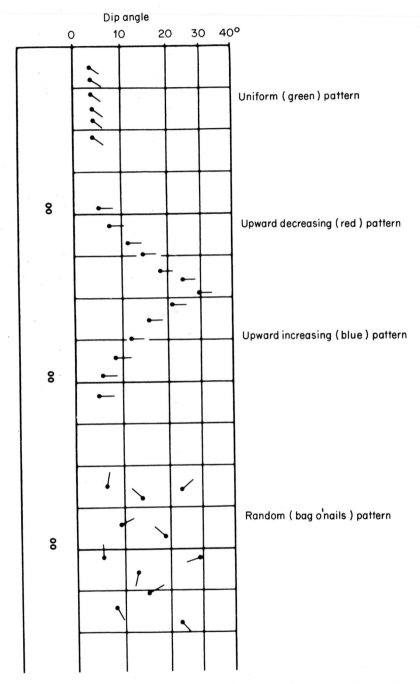

*Figure 1.11*  Dipmeter log in the conventional tadpole plot format. The head of each tadpole marks the amount of dip for a given depth, the tail of the tadpole points in the direction of dip. This figure shows the four basic dipmeter motifs, the significance of which is discussed at appropriate points in the text.

Each environment generates characteristic dip motifs which can be used to determine the geometry and orientation of porous reservoir bodies, such as channels, sand bars and reefs.

When a dipmeter is first run it is usual to calculate dips using a quick and dirty program. This will be able to show however the amount and direction of structural tilt, and to identify faults and unconformities. Obviously structural tilt must be removed from the data before sedimentary dips can be seen. It is thus normal practice to process a dipmeter log at least twice. The second pass will remove tectonic tilt, it will use a program appropriate to the sedimentary fabric present, and it will ultimately display the data both as tadpole plots and in azimuths for carefully defined intervals.

Selection of the appropriate program depends on the type of sedimentary fabric recognized. There are three types of fabric to consider.

*Type I fabric.* This type of fabric occurs in cross-bedded sands. There are generally two sets of planar surface. Major bedding surfaces separate the individual cross-bedded sets (Fig. 1.12). The major bedding surfaces may superficially appear to be horizontal but, depending on their origin, may have a low depositional dip component. Commonly the major bedding surfaces of migrating dunes dip upcurrent at a low angle. In channels the major bedding surfaces of point bars may dip in towards the channel axis. Thus Type I fabrics consist of two populations of planar surfaces, the high angle cross-beds that dip down-current and the subhorizontal set boundaries. Most cross-beds occur in units less than about one metre thick. Thus to examine a Type I fabric the dipmeter log must be processed using a computer program that calculates dips for very small vertical intervals.

*Type II fabric.* This type of fabric is produced by prograding deltas or submarine fans. It therefore consists of a single population of dips, in which the angle of dip generally increases up the increment along with the grainsize (Fig. 1.13). Type II fabrics can be identified from dipmeter computer programs which calculate dips from large vertical intervals.

*Type III fabric.* The third type of fabric has a random arrangement. Planar surfaces may be of short lateral continuity, and random orientation, or the sediment may be completely isotropic. Regardless of the type of program used this fabric can only generate a random pattern of dips (Fig. 1.14). Such random motifs may be due to bad hole conditions or to unsuitable drilling mud. They may however be geologically significant, indicating the presence of slumping, fractures, conglomerates or grain flows.

When the dips have been computed using a suitable program they must be displayed in an appropriate way. The convention is to construct a separate azimuth plot for each genetic increment, defined by the gamma, S.P. or other logs (Fig. 1.15). When azimuths are plotted for dips over large vertical intervals polymodal patterns may result. This is because dips from several increments may have been grouped together. When it will be found that dipmeter logs tend to show four basic motifs on the 'tadpole' plot (Fig. 1.11).

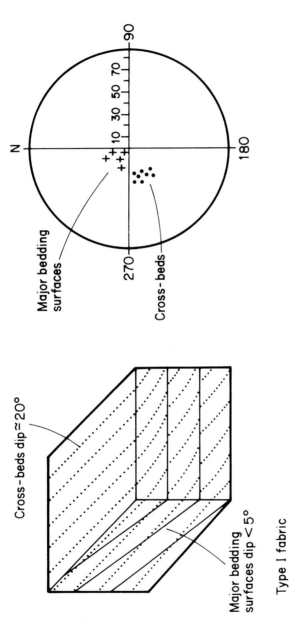

*Figure 1.12* Block diagram and stereo plot to illustrate a Type I fabric, such as occurs in cross-bedded sands.

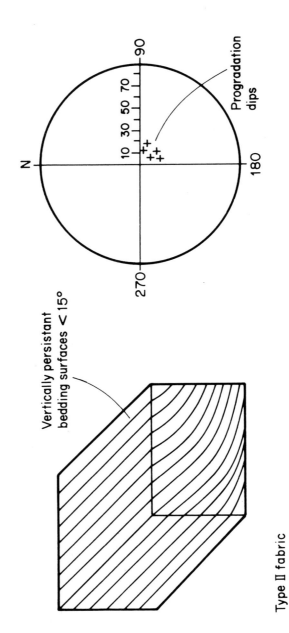

N

10  30  50  70  90

Prograadation
dips

180

270

Vertically persistant
bedding surfaces < 15°

Type II fabric

*Figure 1.13*  Block diagram and stereo plot to illustrate a Type II fabric. This is commonly produced by progradation.

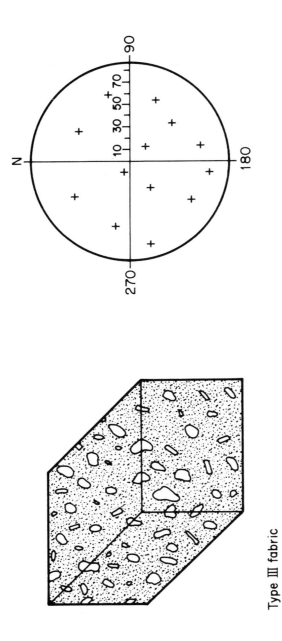

Type Ⅲ fabric

*Figure 1.14* Block diagram and stereo plot to illustrate a Type III fabric. This random fabric may be caused by fractures, conglomerates, slumping or grain flows.

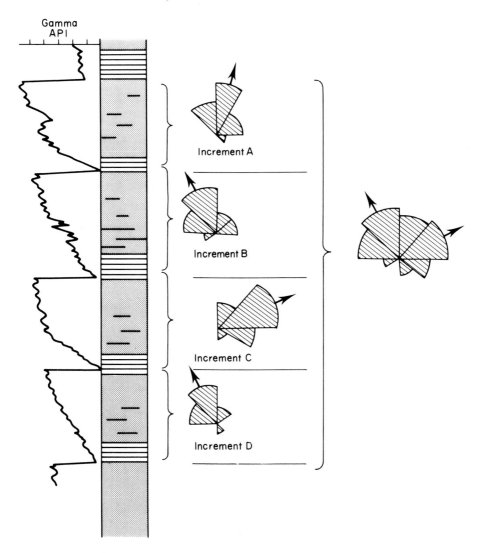

*Figure 1.15* Log to show how dipmeter data plotted increment by increment may show well defined modes. The same data plotted together will yield a polymodal plot.

Steady dips, which for some depth show uniform amounts and directions of dip, are called 'green' patterns. They characterize shale sequences and may be used to determine the structural dip of the formation. Dips of uniform direction but upward decreasing amount are termed 'red' patterns. These are characteristically found in channels, and, as will be shown in subsequent chapters, can be used to locate the trend and axis of the channel. Dips of uniform direction but upward increasing amount are termed 'blue' patterns. These often indicate the direction of progradation of sand bars, delta lobes, submarine fans and reefs.

Lastly the fourth dip motif is all too common. This is aptly termed the 'bag o' nails' or 'drunken tadpole' pattern for intervals whose dips are random, both in amount and direction. This may indicate a poor quality log where, for example, the tool did not work well due to poor borehole conditions, such as caving. 'Bag o' nails' motifs are also characteristic of conglomerates and grainflow sands whose heterogenous fabrics lack planar surfaces.

Further details on the interpretation of sedimentary patterns from dipmeters will be found in Campbell (1968), McDaniel (1968), and Jageler and Matuszak (1972).

In subsequent chapters of this book the various applications of dipmeter interpretation will be discussed in the sub-surface analysis of each environment.

## Summary

This introductory chapter has covered a great deal of ground so it is perhaps appropriate to summarize the preceding review of techniques of environmental interpretation before considering the problems of sub-surface facies analysis.

A sedimentary environment is a part of the earth's surface distinguishable from adjacent areas by physical, chemical, and biological parameters. These areas may be erosional, non-depositional, or depositional. Sedimentary facies originate in depositional environments. They may be defined by their geometry, lithology, sedimentary structures, palaeocurrents, and fossils. These reflect not only the depositional environment but also other earlier and time equivalent environments.

The same kind of sedimentary environments occur repeatedly over the face

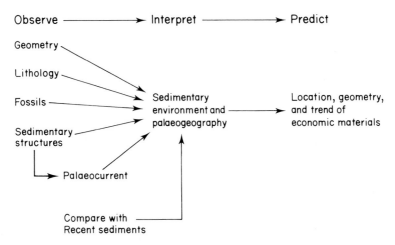

*Figure 1.16*   Illustrative of the basic approach to finding out how a sediment was deposited.

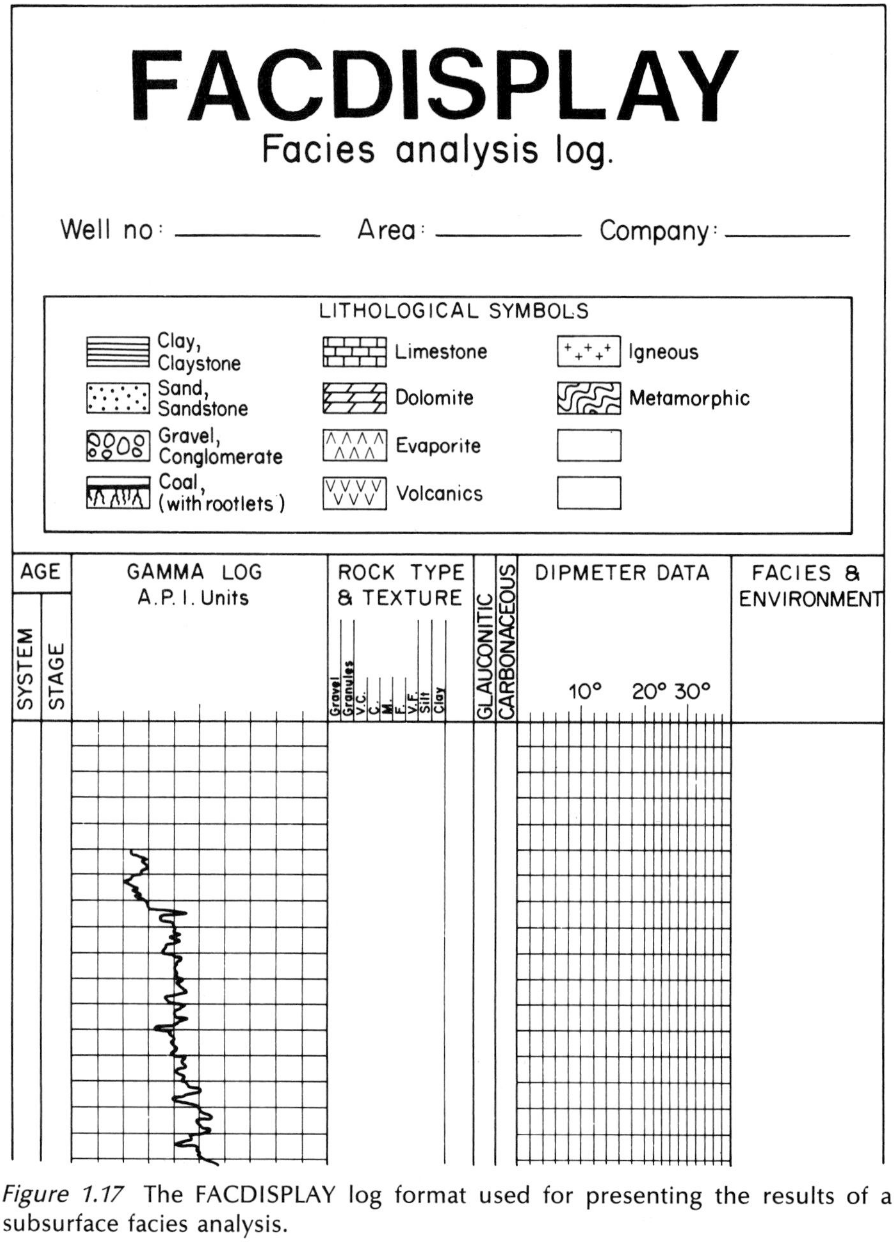
FACDISPLAY
Facies analysis log.
Well no:
Area:
Company:
LITHOLOGICAL SYMBOLS
Clay, Claystone
Sand, Sandstone
Gravel, Conglomerate
Coal, (with rootlets)
Limestone
Dolomite
Evaporite
Volcanics
Igneous
Metamorphic
AGE
SYSTEM
STAGE
GAMMA LOG
A.P.I. Units
ROCK TYPE & TEXTURE
Gravel
Granules
V.C.
C.
M.
F.
V.F.
Silt
Clay
GLAUCONITIC
CARBONACEOUS
DIPMETER DATA
10°
20° 30°
FACIES & ENVIRONMENT

*Figure 1.17* The FACDISPLAY log format used for presenting the results of a subsurface facies analysis.

of the earth today. But no two similar environments are ever identical and the boundaries between different adjacent environments are often transitional.

There appear to be a number of ancient sedimentary facies which occur repeatedly in rocks of different ages all over the earth. These can be related to Recent sedimentary environments.

As environments migrate to-and-fro across the earth they deposit vertical sequences of facies. Thus a conformable vertical sequence of facies can be interpreted according to this concept, known as Walther's Law. Stratigraphic sections can be divided into genetic increments of the same facies type, and may consist of genetic sequences composed of vertically stacked increments of the same origin.

A basic approach to the analysis of ancient sedimentary environments is summarized in Fig. 1.16.

It is generally best to gather all the available data of a sedimentary facies, and not to base the diagnosis of its origin on any single criterion. Geology is at best an imprecise science in which it is seldom possible to make deterministic statements. We deal in probabilities rather than certainties.

The final result of subsurface facies analysis may be illustrated on a FACDISPLAY log format (Fig. 1.17). This presents the gamma and/or S.P. logs, which are the grain-size profile, the dipmeter log, and other relevant data. These are the bare essentials required to present any kind of environmental interpretation.

The student reader must learn to beware of the kind of dogmatic statements with which this chapter abounds. Studious readers may care to immure themselves in the nearest geological library for the next year checking references. Out and out sceptics may prefer to spend the next ten years in the field. Let the rest read on remembering that all generalizations are dangerous including this one.

# REFERENCES

Ager, D. V., 1963. *Principles of Paleoecology*. McGraw-Hill, New York, p. 371.

Allen, J. R. L., 1964. The classification of cross-stratified units with notes on their origin. *Sedimentology*, **2**, 93–114.

——, 1966. On bed forms and palaeocurrents. *Sedimentology*, **6**, 153–90.

——, 1967. Depth indicators of clastic sequences. In: Depth indicators in marine sedimentary environments. (Ed. A. Hallam) *Marine Geology*, Sp. Issue, **5**, No. 5/6, 429–46.

——, 1968. *Current Ripples: their relation to patterns of water and sediment motion*. North Holland Pub. Co., Amsterdam, p. 433.

——, 1970. *Physical processes of sedimentation*. Allen & Unwin, London, p. 248.

——, 1982. *Sedimentary Structures* (2 Volumes). Elsevier, Amsterdam, pp. 593 & 663.

Anstey, N. A., 1982. *Simple Seismics*. I.H.R.D.C., Boston, p. 168.

Asquith, G., 1982. Basic Well Log Analysis for Geologists. *Amer. Assoc. Petrol. Geol. Tulsa*, p. 216.

Bathurst, R. G. C., 1967. Depth indicators in sedimentary carbonates. In: Depth indicators in marine sedimentary environments. (Ed. A. Hallam) *Marine Geology*, Sp. Issue, **5**, No. 5/6, 447–72.

Beerbower, J. R., 1964. Cyclothems and cyclic depositional mechanisms in alluvial plain sedimentation. In: Symposium on Cyclic Sedimentation. (Ed. D. F. Merriam) *Kansas Geol. Surv. Bull*. 169, **1**, 31–42.

Bjerkli, R. & Ostmo-Saeter, J. S., 1973. Formation of glauconite in foraminiferal shells on the continental shelf off Norway. *Mar. Geol.* **14**, 169–78.

Bromley, R. G., 1967. Marine phosphorites as depth indicators. In: Depth indicators in marine sedimentary environment. (Ed. A. Hallam) *Marine Geology*, Sp. Issue, **5**, No. 5/6, 503–10.

Busch, D. A., 1971. Genetic units in delta prospecting. *Bull. Amer. Assoc. Petrol. Geol.*, **55**, 1137–1154.

Campbell, R. L., 1968. Stratigraphic Applications of Dipmeter Data in Mid continent. *Bull. Amer. Assoc. Petrol. Geol.*, **52**, 1700–19.

Chayes, F., 1956. *Petrographic Modal Analysis. An elementary statistical appraisal.* Wiley & Sons, New York, p. 113.

Coffeen, J. H., 1978. *Seismic Exploration Fundamentals.* Petroleum Publishing Corp., Tulsa, p. 277.

Collinson, J. D. and Thompson, D. B., 1982. *Sedimentary Structures.* Allen & Unwin, London, p. 240.

Conybeare, C. E. B., 1976. *Geomorphology of oil and gas fields in sandstone bodies.* Elsevier, Amsterdam, p. 341.

——, and Crook, K. A. W., 1968. Manual of sedimentary structures. *Bureau of Mineral Resources.* Bull. No. 102, Canberra A.C.T., p. 327.

Crimes, T. P., and Harper, J. C. (Eds) *Trace fossils.* Liverpool 1970. Geol. Soc., p. 547.

——, and Harper, J. C., 1977. *Trace Fossils.* Volume II. Seel House Press, Liverpool, p. 351.

Crosby, E. J., 1972. Classification of sedimentary environments. In: Recognition of Ancient sedimentary environments. *Soc. Econ. Pal. Min.*, Sp. Pub., No. 16, pp. 1–11.

Cummings, W. A., 1962. The greywacke problem. *Lpool. Manchr. Geol. J.*, **3**, 51–72.

Degens, E. T., 1965. *Geochemistry of Sediments: a brief survey.* Prentice-Hall, N.J., p. 342.

——, Williams, E. G., and Keith, M. L., 1957. Environmental studies of Carboniferous sediments. In: Geochemical criteria for differentiating marine and freshwater shales. *Bull. Amer. Assoc. Petrol. Geol.*, **42**, 981–97.

Dodd, J. R. and Stanton, R. J., 1981. *Palaeoecology, Concepts and Applications.* J. Wiley, London, p. 544.

Doeglas, D. J., 1946. Interpretation of the results of mechanical analysis. *J. Sediment.* Elsevier, Amsterdam, p. 280.

Duff, P. McL. D., Hallam, A., and Walton, E. K., 1967. *Cyclic Sedimentation.* Elsevier, Amsterdam, p. 280.

Dunham, R. J., 1962. Classification of carbonate rocks according to their depositional texture. In: Classification of carbonate rocks. (Ed. W. E. Ham) *Amer. Assoc. Petrol. Geol.* Memoir No. 1, pp. 108–21.

Feldhausen, P. H., and Ali, S. A., 1975. Sedimentary facies of Barataria Bay, Louisiana, determined by multivariate statistical techniques. *Sedimentary Geology*, **14**, 259–74.

Fisher, W. L., 1969. Facies characterization of Gulf Coast basin delta systems, with some Holocene analogues. *Gulf Coast Association Geol. Soc. Trans.* **19**, 239–261.

Folk, R. L., 1967. A review of grain size parameters. *Sedimentology*, **6**, 73–94.

Frey, R. W. (Ed.) 1975. *The study of trace fossils.* Springer-Verlag, Berlin, p. 562.

Gall, J. C., 1983. *Ancient Sedimentary Environments and the Habitats of Living Organisms.* Springer-Verlag, Berlin, p. 240.

Galloway, W. E., 1968. Depositional systems of the Lower Wilcox Group, North Central Gulf Coast Basin. *Gulf Coast Association Geol. Soc. Trans.* **18**, 275–89.

Gecker, R. F., 1957. Introduction to Palaeoecology. French translation, *Bases de la Palaeoecologie*. Bur. Récherches Geol. Min., p. 83, also, spelt Hecker, 1965. American Elsevier Pub. Co. N. Yk., p. 166.

Ginsburg, R. N., Lloyd, R. M., Stockman, K. W., and McCallum, J. S., 1963. Shallow water carbonate sediments. In: *The Sea*, vol. III. (Ed. M. N. Hill) Wiley & Sons, N. Yk. pp. 554–82.

Grim, R. E., 1958. Concept of diagenesis in argillaceous sediments. *Bull. Amer. Assoc. Petrol. Geol.*, **42**, 246–53.

Gubler, Y. (Ed.) 1966. *Essai de nomenclature et caractérisation des principales structures sédimentaries*. Ed. Tecnip. Paris, p. 29.

Harms, J. C., and Tackenberg, P., 1972. Seismic signatures of sedimentation models. *Geophysics*, **37**, 45–58.

Heckel, P. H., 1972. Recognition of Ancient shallow marine environments. In: Recognition of Ancient sedimentary environments, (Ed. J. K. Rigby and W. K. Hamblin) *Soc. Econ. Pal. Min.*, Sp. Pub., No. 16. pp. 226–86.

Hedgpeth, J. W., and Ladd, H. S. (Eds) 1957. Treatise on marine ecology and paleoecology, *Mem. Geol. Soc. Amer.* No. 67 (2 vols.) pp. 1296 and 1077.

Howard, J. D., 1975. The sedimentological significance of trace fossils. In: *The study of trace fossils*. (Ed. R. W. Frey) Springer-Verlag, New York, pp. 131–46.

Imbrie, J., and Newell, N. D. (Eds) 1964. *Approaches to Paleoecology*. Wiley & Sons, N.Y. p. 432.

——, and Purdy, E. G., 1962. Classification of modern carbonate sediments. In: Classification of carbonate rocks. (Ed. W. E. Ham) *Amer. Assoc. Petrol. Geol.* Memoir No. 1, pp. 253–73.

Ireland, B. J., Curtis, C. D., and J. A. Whiteman, 1983. Compositional variation within some glauconites and illites and implications for their stability and origins. *Sedimentology*, **30**, 769–86.

Jageler, A. H., and Matuszak, D. R., 1972. Use of well logs and dipmeters in stratigraphic trap exploration. In: Stratigraphic oil and gas fields. (Ed. R. E. King) *Amer. Assoc. Pet. Geol.* Spec. Pub. No. 10, pp. 107–35.

Jopling, A. V., and Walker, R. G., 1967. Morphology and origin of ripple drift cross-lamination, with examples from the Pleistocene of Massachusetts. *J. Sediment. Petrol.*, **38**, 971–84.

Klein, G. de V., 1967. Paleocurrent analysis in relation to modern marine sediment dispersal patterns. *Bull. Amer. Assoc. Petrol. Geol.*, **51**, 366–82.

Kleyn, A. H., 1982. *Seismic Reflection Interpretation*. Applied Science Publishers, Barking, p. 269.

Krinsley, D., and Cavallero, L., 1970. Scanning electron microscope examination of preglacial eolian sands from Long Island, New York. *J. Sediment Petrol.*, **40**, 1345–50.

——, and Doornkamp, J. C., 1973. *Atlas of quartz sand surface textures*. Cambridge Univ. Press, p. 91.

——, Friend, P. F., and Klimentidis, R., 1976. Eolian transport textures on the surface of sand grains of Early Triassic Age. *Bull. Geol. Soc. Amer.*, **87**, 130–2.

——, and Funnel, B. M., 1965. Environmental history of sand grains from the Lower and Middle Pleistocene of Norfolk, England. *Quart. J. Geol. Soc. London.*, **121**, 435–62.

Krumbein, W. C., and Sloss, L. L., 1959. *Principles of Stratigraphy and Sedimentation*. Freeman, San. Fran., p. 497.

Kuenen, P. H., 1960. Experimental abrasion of sand grains. *Rept. Int. Geol. Cong.* 21 Session, Pt. 10. Submarine Geology.

Logvinenko, N. V., 1982. Origin of glauconite in the Recent bottom sediments of the ocean. *Sed. Geol.*, **31**, 43–48.

Lyons, P. L., and Dobrin, M. B., 1972. Seismic exploration of stratigraphic traps. In: Stratigraphic Oil and Gas Fields (Ed. R. E. King) *Amer. Assoc. Pet. Geol.*, No. 16, pp. 225–43.

McCall, P. L. and Tevesz, M. J. S., 1982. *Animal-Sediment Relations*. Plenum Press, New York, p. 366.

McDaniel, G. A., 1968. Application of sedimentary directional features and scalar properties to Hydrocarbon Exploration. *Bull. Amer. Assoc. Petrol. Geol.*, **52**, 1689–99.

Merriam, D. F. (Ed.) 1967. *Computer Applications in the Earth Sciences. Colloquium on time-series analysis. Computer contribution No. 18. State Geological Survey.* University of Kansas.

Middleton, G. V., 1973. Johannes Walther's law of correlation of facies. *Bull. Geol. Soc. Am.*, **84**, pp. 979–88.

Mills, P. C., 1983. Genesis and diagnostic value of soft sediment deformation structures—a review. *Sed. Geol.*, **35**, 83–104.

Moiola, R. J. and Weiser, D., 1968. Textural parameters: an evaluation. *J. Sediment Petrol.*, **38**, 45–53.

Murty, V. N., 1975. Palaeosalinity studies of some PreCambrian shales of the Bohai-Keonijhar Belt, Orissa—a preliminary appraisal. *Jour. Geol. Soc. India*, **16**, 230–4.

Odin, G. S. and Mather, A., 1981. De glauconarium origine. *Sedimentology*, **28**, 611–41.

Pettijohn, F. J., and Potter, P. E., 1964. *Atlas and Glossary of Primary Sedimentary Structures*. Springer-Verlag, N.Y., p. 370.

——, and Siever, P. E., 1972. *Sand and Sandstone*. Springer-Verlag, New York, p. 618.

Pirson, S. J., 1983. *Geologic well log analysis* (3rd Edn.) Gulf Pub. Corp., Houston, p. 424.

Porrenga, D. H., 1967. Glauconite and chamosite as depth indicators in the marine environment. In: Depth indicators in marine sedimentary rocks. (Ed. A. Hallam) *Marine Geology*, Sp. Issue **5**, No. 5/6, pp. 495–502.

Potter, P. E., 1967. Sand bodies and sedimentary environments: a review. *Bull. Amer. Assoc. Petrol. Geol.*, **51**, 337–65.

——, and Pettijohn, F. J., 1977. *Paleocurrents and Basin Analysis*. (2nd Edn.) Springer-Verlag, Berlin, p. 425.

——, Shimp, N. F., and Witters, J., 1963. Trace elements in marine and freshwater argillaceous sediments. *Geochem. Cosmochim. Acta*, **27**, 669–94.

Raudkivi, A. J., 1976. *Loose Boundary hydraulics*. Pergamon Press, London, p. 397.

Payton, C. E., 1977. Seismic Stratigraphy—applications to hydrocarbon exploration. Mem. No. 26. *Amer. Assoc. Petrol. Geol. Tulsa*, p. 516.

Pryor, W. A., 1973. Permeability—porosity patterns and variations in some Holocene Sand Bodies. *Bull American Association Petrol. Geol.*, **57**, 162–89.

Reed, W. A., Le Fever, R., and Moir, G. J., 1975. Depositional environment from settling-velocity (Psi) distributions. *Bull Geol. Soc. Amer.*, **86**, 1321–8.

Reineck, H. E., and Singh, I. B., 1973. *Depositional sedimentary environments*. Springer-Verlag, New York, p. 471.

Rodriguez, J., and Gutschick, R. C., 1970. Late Devonian-Early Mississippian Ichnofossils from western Montana and northern Utah. In: *Trace fossils*. (Ed. T. P. Crimes, and J. C. Harper) Liverpool Geol. Soc., pp. 407–438.

Rohrlich, V., Price, N. B., and Calvert, S. J., 1969. Chamosite in the Recent Sediments of Loch Etive, Scotland. *J. Sediment. Petrol.* **39**, 624–631.

Schwarzacher, W., 1975. *Sedimentation Models and Quantitative Stratigraphy*. Elsevier, Amsterdam, p. 396.

Seilacher, A., 1967. Bathymetry of trace fossils. In: Depth indicators in marine sedimentary environments. (Ed. A. Hallam) *Marine Geology*, Sp. Issue, **5**, No. 5/6, pp. 413–28.

Selley, R. C., 1968. A classification of paleocurrents models. *J. Geol.*, **76**, 99–110.

——, 1969a. Torridonian alluvium and quicksands. *Scott. J. Geol.*, **5**, 328–46.

——, 1969b. Studies of sequence in sediments using a simple mathematical device. *Quart. J. Geol. Soc. Lond.* **125**, 557–81.

——, 1982. *Introduction to Sedimentology*. (2nd Edn) Academic Press, London, p. 419.

——, 1976. Sub-surface environmental analysis of North Sea sediments. *Bull. Amer. Assoc. Pet. Geol.*, **60**, 184–95.

Shaw, H. F., 1980. *Clay Minerals in Sediments and Sedimentary Rocks. Developments in Petroleum Geology* Volume 2. (Ed. G. D. Hobson) Applied Science Pubs, Barking, pp. 53–86.

Shelton, J. W., 1967. Stratigraphic models and general criteria for recognition of alluvial, barrier-bar, and turbidity current sand deposits. *Bull. Amer. Assoc. Petrol. Geol.* **51**, 2441–60.

Shepard, P. E., 1964. Criteria in modern sediments useful in recognizing ancient sedimentary environments. In: *Deltaic and Shallow Marine Sediments*. (Ed. L. M. J. U. Van Straaten) Elsevier, Amsterdam, pp. 1–25.

Sheriff, R. E., 1976. Inferring stratigraphy from seismic data. *Amer. Assoc. Pet. Geol. Bull.*, **60**, 528–42.

Shimp, N. F., Witters, J., Potter, P. E., and Schleicher, J. A., 1969. Distinguishing marine and freshwater muds. *J. Geol.*, **77**, 566–80.

Simons, D. B., Richardson, E. V., and Nordin, C. F., 1965. Sedimentary structures generated by flow in alluvial channels. In: Primary sedimentary structures and their hydrodynamic significance. (Ed. G. V. Middleton) *Soc. Econ. Min. Pal.*, Sp. Pub. No. 12, pp. 34–42.

Tanner, W. F., 1967. Ripple mark indices and their uses. *Sedimentology*, **9**, 89–104.

Vail, P. R., Mitchum, R. M., Todd, R. G., Widmier, J. M., Thompson, S., Sangree, J. B., Bubb, J. N. and Hatledid, W. G., 1977. Seismic Stratigraphy and Global Changes in Sea Level. In: Seismic Stratigraphy — applications to Hydrocarbon Exploration. (Ed. C. E. Payton) *Amer. Assoc. Petrol. Geol. Mem. 26*, pp. 49–212.

Visher, G. S., 1965. Use of vertical profile in environmental reconstruction. *Bull. Amer. Assoc. Petrol. Geol.*, **49**, 41–61.

——, 1969. Grain-size distributions and depositional processes. *J. Sedim. Pet.*, **39**, 1074–106.

Walker, R. G., 1963. Distinctive types of ripple drift cross-lamination. *Sedimentology*, **2**, 173–88.

Walther, J., 1893. *Einleithung in die Geologie als Historiche Wissenschaft Band 1. Beobachtungen uber die Bildung der Gesteine und ihter organischen Einschlusse.* G. Fischer, Jena, p. 196.

Weaver, C. E., 1958. Geologic interpretation of argillaceous sediments. Part I. Origin and significance of clay minerals in sedimentary rocks. *Bull. Amer. Assoc. Petrol. Geol.*, **42**, 254–71.

White, W. R., Milli, H., and Crabbe, A. D., 1976. Sediment transport theories: a review. *Proc. Inst. Civil. Eng.*, Part 2, pp. 265–93.

# 2 River deposits

## INTRODUCTION: RECENT ALLUVIUM

Thick sequences of red-coloured interbedded conglomerates, sands, and shales devoid of marine fossils are common all over the world. These are generally believed to be alluvial in origin. The geomorphology, hydrology, and sedimentology of Recent rivers are well known from both observational and experimental studies (see Chorley (1969), Gregory (1977), Schumm (1977) and Collinson and Lewin (1983) for reviews).

Recent alluvial deposits are basically of two kinds whose characteristics are largely related to the morphology of the river channels which deposited them:

(1) High-sinuosity meandering channels
(2) low-sinuosity braided channel complexes.

### Alluvium of meandering rivers

High-sinuosity meandering river channels typically develop where gradients and discharge are relatively low compared to those of braided channel systems (Schumm, 1981 and Cant, 1982). At the present time they are characteristic of humid vegetated parts of the world where seasonal discharge rates are fairly steady and sediment availability is relatively low due to subdued topographic relief and the impeding effect of vegetation both on soil erosion and lateral erosion of channel margins. The Mississippi is a good example of a Recent meandering river (Turnbull, Krinitsky, and Johnson, 1950; Fisk, 1944).

The deposits of a meandering river can be sub-divided into three main sub-facies due to deposition in three different sub-environments (Shanster, 1951).

#### Floodplain sub-facies

Sheets of very fine sand, silt and clay are deposited on the overbank areas of the river's floodplain. These are laminated, ripple-marked, and often contain horizons of sand-filled shrinkage cracks, suggesting sub-aerial exposure. The presence of soils may be indicated by carbonate caliches, ferruginous laterites, and rootlet horizons. Peat may form and detrital plant debris be preserved on bedding surfaces. This sub-facies is deposited largely out of suspension during floods when the river breaks its banks.

#### Channel sub-facies

The lateral migration of a meandering channel erodes the outer concave bank, scours the river bed, and deposits sediment on the inner bank (point bar). This produces a characteristic sequence of grain size and sedimentary structures. At the base is an erosion surface overlain by extraformational pebbles,

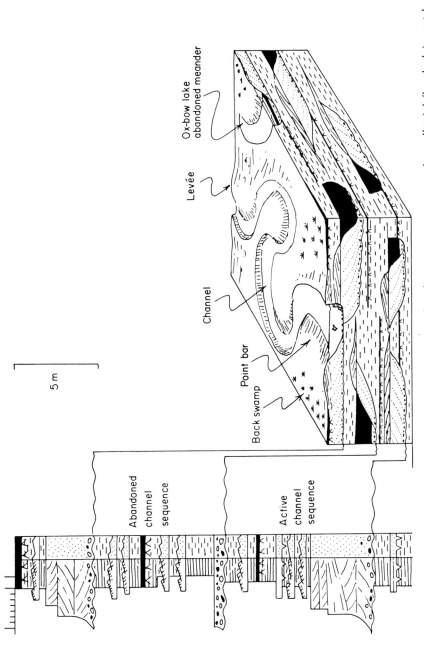

*Figure 2.1* Geophantsmogram to illustrate the subenvironments and sedimentary sequence of an alluvial flood plain cut by a meandering river channel (from Selley, 1982 courtesy of Academic Press).

Ox-bow lake abandoned meander

Levée

Channel

Point bar

Back swamp

5 m

Abandoned channel sequence

Active channel sequence

intraformational mud pellets, fragmented bones, and waterlogged drift wood. These originated as a lag deposit on the channel floor and are overlain by a sequence of sands with a general vertical decrease in grain size. Massive, flat-bedded and trough cross-bedded sands grade up into tabular planar cross-bedded sands of diminishing set height. These in turn pass up into micro-crosslaminated and flat-bedded fine sands which grade into silts of the floodplain sub-facies (Mackin, 1937; Allen, 1964; Visher, 1965).

### Abandoned channel sub-facies
Curved shoestrings of fine-grained deposits infill abandoned channels, sometimes called ox-bow lakes, formed when the river meanders back on itself short-circuiting the flow. This sub-facies is similar to floodplain deposits but is distinguishable from them by its geometry and because it abruptly overlies a channel lag conglomerate with no intervening point bar sand sequence.

The origin and geometry of the sub-facies deposited by meandering rivers are summarized in Fig. 2.1

## Alluvium of braided rivers
Braided river systems consist of an interlaced network of low sinuosity channels. Recent examples occur on steeper gradients and have higher discharges than meandering rivers (Miall, 1977; Rust, 1979; Rachocki, 1981; Schumm, 1981; Cant, 1982 and Nilson, 1982). Many present-day braided rivers are found on piedmont fans at the edges of mountains where there are large amounts of sediment and discharge is often, but not always, seasonal. Examples have been described both from the hot deserts of the U.S.A., Algeria and Australia (Blissenbach, 1954; Williams, 1970 and 1971 respectively), and also from periglacial mountainous regions such as Western Canada (Smith, 1983), and the Yukon (Williams and Rust, 1969; Rust, 1972). In these regions erosion is rapid, discharge is sporadic and high, and there is little vegetation to hinder runoff. Because of these factors rivers are generally overloaded with sediment. A channel is no sooner cut than it chokes in its own detritus. This is dumped as bars in the centre of the channel around which two new channels are diverted (Knight, 1975). Repeated bar formation and channel branching generates a network of braided channels over the whole depositional area. Thus the alluvium of braided rivers is typically composed of sand and gravel channel deposits to the exclusion of fine-grained overbank silts and clays. Due to repeated channel switching and fluctuating discharge there is generally an absence of laterally extensive cyclic sequences similar to those produced by meandering channels. Fining-upward gravel, sand, silt sequences have been recorded however and are attributed to waning current velocities as a channel is gradually infilled (William and Rust, 1969). Coarsening-upward sequences have been described from the Pleistocene glacial outwash sands and gravels of southern Ontario (Costello and Walker, 1972). These are attributed to the infilling of braided channels at times of rising flood stages.

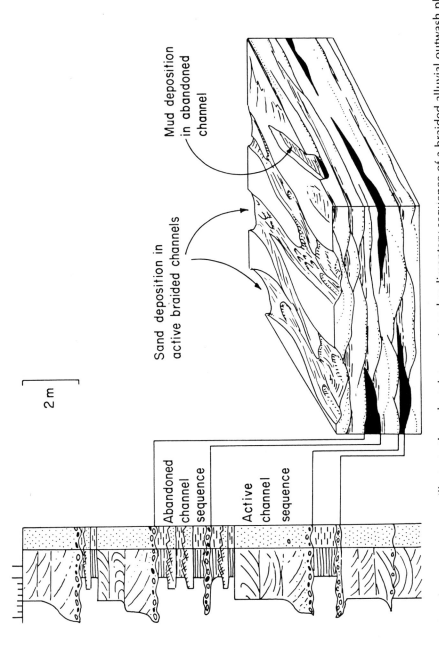

2 m

Mud deposition
in abandoned
channel

Sand deposition in
active braided channels

Abandoned
channel
sequence

Active
channel
sequence

*Figure 2.2* Geophantasmogram to illustrate the subenvironments and sedimentary sequence of a braided alluvial outwash plain (from Selley, 1982 courtesy of Academic Press).

The small amount of silt which is present in braided alluvium is generally deposited in abandoned channels. These form both by channel choking and switching (Doeglas, 1962) and by river piracy due to rapid headward erosion of channels on the downslope part of a fan (Denny, 1967, Fig. 1). Where an abandoned channel still connects to an active channel downstream it can form a trap for silt and clay carried into it by reverse eddies from the main channel. Thus fine-grained sediments form with shoestring geometries which correspond to the original channel form.

It is apparent from this review of processes in modern braided channel systems that, though both upward-fining and upward-coarsening sequences of grain size

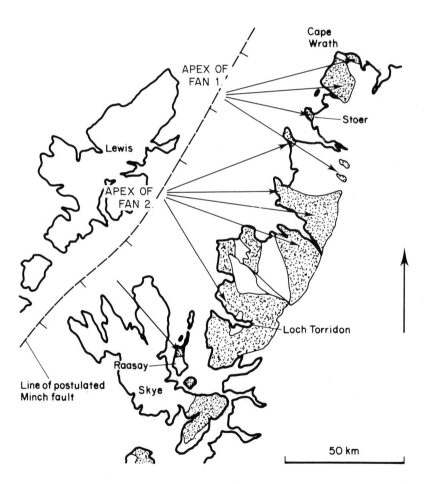

*Figure 2.3* Distribution of Torridonian (PreCambrian) sediments, northwest Scotland. Arrows indicate red facies palaeocurrents. Note how these describe two radiating fans whose apices lie along the Minch fault. Modified from Williams, 1969, Fig. 12; by courtesy of the *Journal of Geology*.

may be formed sometimes, the gross arrangement of sub-facies tends to be haphazard. The resultant formation tends to be made up of coalesced braided channel sands and gravels with rare abandoned channel shale shoe-strings (Fig. 2.2). This is in marked contrast to the more regular arrangement of sub-facies deposited by meandering channel systems, with their laterally extensive overbank shales.

Two case histories of ancient alluvium will now be described. In the first of these emphasis will be placed on the deposits of braided rivers, in the second on those of meandering channels.

## The Torridon Group (Precambrian) of northwest Scotland: description and discussion

The Torridonian rocks crop out along the northwest coast of Scotland and on adjacent islands of the Hebrides (Fig. 2.3). These are fascinating rocks which have been attributed to nearly every depositional environment and climate known on earth (see Van Houten, 1961). An extraterrestrial origin was proposed by Parkes (1963) who suggested that the Torridonian dropped off the moon. First the facts: these sediments overlie the Lewisian gneiss with an irregular unconformity and are themselves separated by a planar unconformity from overlying Lower Cambrian sandstones and limestones.

The Torridonian is divisible into two groups by an unconformity which separates westerly dipping red sandstones and shales of the Stoer Group from the overlying sub-horizontal Torridon Group. This consists essentially of about 3 km of interbedded red conglomerates, sandstones, and shales. In the north these deposits are separated from the Lewisian gneiss by a planar pre-Torridonian weathered surface. Traced southward, they overlie the Stoer Group. Still further south around the type area of Loch Torridon these deposits overlie and infill a dissected topography cut into the gneiss basement with a relief of several hundred feet. In low-lying parts of this surface breccias banked against the flanks of pre-Torridonian mountain pass laterally into grey shales and sandstones.

Such a situation can be seen on the island of Raasay whose Torridon Group sediments will now be described. These are divisible into three major facies defined as follows:

(i) *Basal facies:* Red and grey breccias and granulestones present adjacent to the buried gneiss topography.

(ii) *Grey facies:* Grey sandstones and shales, laterally equivalent to the basal facies.

(iii) *Red facies:* Coarse red pebbly sandstones and siltstones overlie the previous two facies.

The regional geometries of the three facies are summarized in Fig. 2.4. Each facies will now be described and its environment discussed.

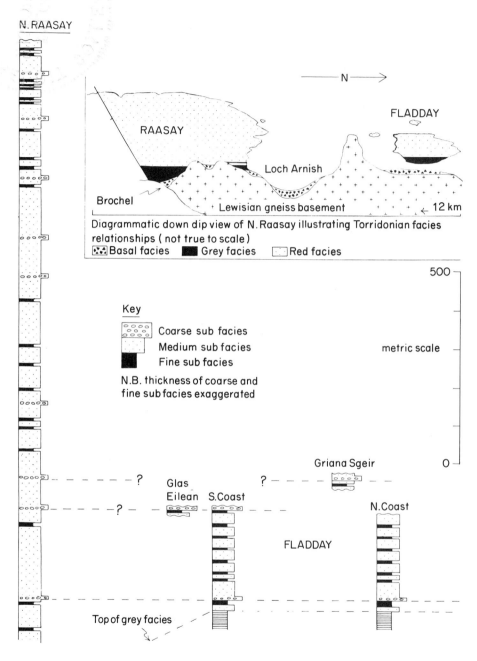

*Figure 2.4* Measured sections and facies distribution of Torridon Group sediments on Raasay. Reproduced from Selley, 1969, Fig. 1; by courtesy of the *Scottish Journal of Geology*.

*Figure 2.5* Lewisian gneiss boulder in basal facies conglomerate, west coast of Raasay opposite Fladday island. From Selley, 1965, Fig. 6; *Journal of Sedimentary Petrology*; by courtesy of the Society of Economic Paleontologists and Mineralogists.

## Basal facies: description

On Raasay the sub-Torridonian unconformity (after allowing for later tectonic tilting) is a dissected plateau with a steep westerly facing scarp some 60 m high. This topography has largely been stripped of its Torridonian cover.

Basal facies deposits can still be found, however, infilling gullies on the scarp crest and as fan-shaped deposits banked against its lower slope.

These sediments are boulder beds, breccias, and granulestones. Individual boulders are up to 2 m in diameter (Fig. 2.5). The composition of the boulders and the arkosic nature of the granulestones leaves no doubt that they are locally derived from the Lewisian gneiss. These sediments are extremely poorly sorted, generally massive, or rudely bedded.

When allowance is made for regional tectonic tilting (by correction using a stereographic net) it can be seen that the bedding of this facies has a considerable depositional dip. In the basal facies on the east side of Loch Arnish dips radiate from exhumed valleys cut in the Lewisian gneiss (Fig. 2.6). These, the stratigraphically lowest basal facies deposits, are red in colour due to a ferric oxide matrix. Higher up the succession, where they pass laterally into the grey facies, the basal facies themselves have a grey-green chloritic cement. Higher still where the basal facies infill gullies on the crest of the Lewisian plateau they are red, like the red facies with which they are laterally equivalent.

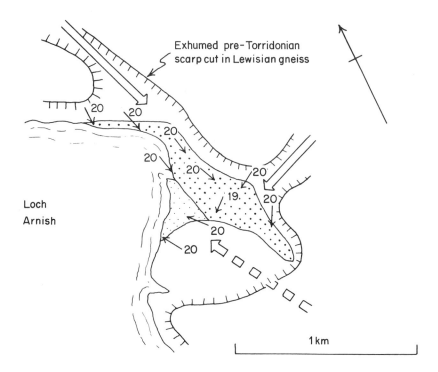

*Figure 2.6* Distribution of red basal facies of the Torran outlier, north Raasay. Conglomerates (large dots) overlain by pebbly granulestones (small dots). Small arrows and numbers indicate orientation and inclination of depositional dip, after allowing for tectonic tilting. Thick arrows indicate presumed direction of scree movement down pre-Torridonian valleys cut in the Lewisian gneiss. Conspicuous parent valley for upper granulestones absent, but finer grain size suggests a more distant source. From Selley, 1965, Fig. 5, *Journal of Sedimentary Petrology*; by courtesy of the Society of Economic Paleontologists and Mineralogists.

## Basal facies: interpretation

The coarse grain size, angularity, poor sorting, and petrography of the basal facies deposits clearly show that they were derived locally from the Lewisian gneiss. Their geometry and the radiating depositional dips leaves little doubt that they are ancient piedmont fans. Deposition was probably due to avalanches, mudflows, and sheet floods such as occur on steep slopes near the apices of Recent fans.

## Grey facies: description

The basal facies pass laterally into the grey facies at Brochel and along the Raasay coast opposite the island of Fladday. The grey facies consist of over 130 m of

*Figure 2.7* Interbedded graded greywackes (?turbidites) and shales with desiccation cracks, overlain by coarse cross-bedded channel sand. Grey facies; Brochel foreshore, east Raasay. From Selley, 1965, Fig. 12. *Journal of Sedimentary Petrology*; by courtesy of the Society of Economic Paleontologists and Mineralogists.

three interbedded rock types: thick beds of coarse sandstone, thin beds of medium and fine sandstone, and shales.

Shales are the most abundant rock type, composing about 80% of the grey facies section at Fladday and about 50% at Brochel. These are poorly sorted fissile siltstones, grey in colour, laminated, argillaceous, and micaceous. Thin clay and sand laminae are frequent. Within these shales are lenses and isolated ripples of very fine micro-crosslaminated sands. Desiccation cracks infilled by red medium well-rounded (? wind-blown) sands are common.

Interbedded with the shales are 10–15-cm thick medium and fine-grained sandstones. These are of two types. In the Fladday section are cleanwashed arkoses, micro-crosslaminated from top to bottom. At Brochel, however, equivalent sands are largely poorly sorted vertically graded greywackes. Their bases are often conspicuously erosional and overlain by a thin layer of granules and shale clasts. Internally these sands are generally massive or laminated with occasional convolutions. In contrast to the equivalent beds on Fladday micro-crosslamination is restricted to the top two or three centimetres of these units (Fig. 2.7).

Interbedded with the silts and greywackes at Brochel are coarse grey cross-bedded arkose channel sandstones (Fig. 2.7). Apart from their drab colour these are similar to those of the red facies above.

## Grey facies: interpretation

The abundance of laminated grey shale suggests that this facies originated in a low energy environment where fine-grained sediment settled out of suspension. The presence of desiccation cracks shows that the water was shallow and occasionally receded or evaporated.

Intermittent high-energy conditions are indicated by the beds of medium and fine sandstone. The well-sorted, micro-crosslaminated sands of Fladday were probably deposited by gentle traction currents. The poorly-sorted graded greywackes at Brochel, however, show many of the typical features of turbidites (see p. 246).

The coarse cross-bedded channel sands at Brochel herald the advance of the overlying red facies alluvium and, as they debouched into the waters of the grey facies, perhaps generated the turbidites.

Considered overall therefore the grey facies was deposited mostly under quiet shallow water which from time to time receded or evaporated.

Sands, deposited both by traction currents and turbidity flows, indicate higher-energy conditions which ultimately dominated the area due to burial by the red facies. One of the problems of the grey facies, however, is whether it was deposited within enclosed lake basins or in the bays of a dissected coastline. For Phanerozoic rocks this problem would be quickly solved by palaeontology. Though the grey facies shales do contain microfossils their environmental significance is uncertain. Likewise the sedimentological evidence is equivocal. Desiccation cracks can form on both tidal flats and dried out lakes. Turbidity flows have been recorded both from Recent lakes and fjords (p. 249). Palaeocurrents in the grey facies are unipolar, but even bipolar palaeocurrents would not be significant since they have been recorded from ancient lakes and are not therefore exclusive to marine tidal deposits (p. 107). Outcrop is not good enough to show whether the grey facies are restricted to isolated hollows in the gneiss basement or whether they are mutually connected along a dissected shoreline.

Geochemical data support a non-marine origin, the boron content of illites in the shales being in the order of 100 ppm, whereas illites from marine shales commonly show more than twice that amount (Stewart and Parker, 1979).

## Red facies: description

The grey and basal facies of the Torridon Group are quantitatively insignificant. The bulk of the sediments are of the red facies which is some 2 km thick on Raasay and over 3 km on the Scottish mainland. The red facies is composed largely of coarse red pebbly arkoses and rare red shales. There are considerable lateral and vertical variations in grain size both on Raasay and adjacent islands and on the mainland.

The red facies on Raasay can be divided into coarse-, medium-, and fine-grained sub-facies defined by their grain size and sedimentary structures.

*Figure 2.8* Ruptured quicksand in red facies sandstone. Foreshore, northwest coast of Fladday island. Reproduced from Selley and others, 1963, plate XVI. Fig. 1; by courtesy of the *Geological Magazine*.

The coarse sub-facies is composed of very coarse red pebbly arkoses. It occurs in sheets, seldom more than a few metres thick, which can be traced laterally across the southeastwards palaeoslope for kilometres (Fig. 2.4). The bases of these sheets are undulose erosion surfaces overlain by thin intra- and extra-formational conglomerates. These pass up into very coarse pebbly sandstones which show flat-bedding and both trough and tabular planar cross-bedding. Penecontemporaneous deformation can often be seen, including recumbent foresets, convolute bedding, and huge diapiric structures several metres high (Fig. 2.8).

The medium-grained sub-facies makes up the bulk of the red facies section on Raasay, being composed of coarse-, medium-, and fine-grained arkoses. These generally occur in lenticular beds about a metre thick; bounded above and below by red siltstone layers one or two centimetres thick. These sandstones are better sorted than those of the previous sub-facies and pebbles are generally absent. Sedimentary structures include tabular planar cross-bedding and horizontal bedding. These are often deformed in the various manners described for the previous sub-facies. Layers of heavy minerals, mainly iron ore, are common in this and the previous facies. They too are often deformed.

The fine sub-facies makes up about 10% of the total section of the red facies. It is composed of dark red soft weathering shaley argillaceous siltstone with thin interbeds of very fine pink sandstones. The sandstones are often deformed

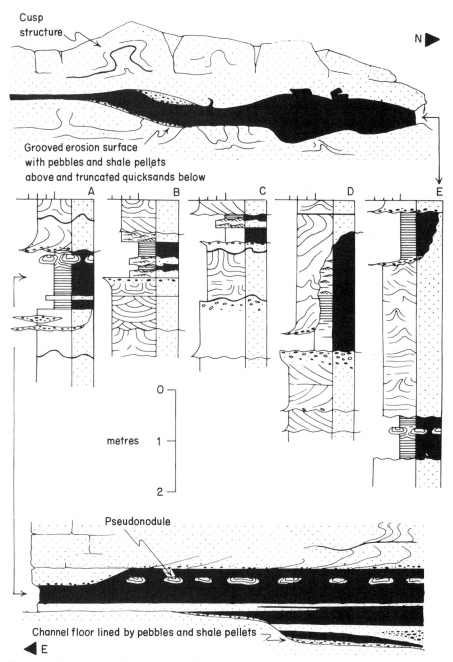

Cusp
structure

N ▶

Grooved erosion surface
with pebbles and shale pellets
above and truncated quicksands below

A          B          C          D          E

0

metres    1

2

Pseudonodule

Channel floor lined by pebbles and shale pellets

◀ E

*Figure 2.9*  Measured sections and sketches of silt-infilled abandoned channels in red facies sandstones, Raasay. From Selley, 1969, Fig. 4; by courtesy of the *Scottish Journal of Geology.*

due to penecontemporaneous collapse into the underlying shale forming loadcasts and pseudo-nodules. Internally the sands are generally micro-cross-laminated throughout, with ripples preserved on the upper surface where they are overlain by siltstone. The siltstones are generally laminated and sometimes contain laminae and isolated ripples of sand.

The fine-grained sub-facies occurs in sequences about a metre thick. At the base they overlie laterally extensive scoured surfaces which truncate primary bedding and quicksand structures of the sand beneath. The contact is marked by a thin conglomerate of extraformational pebbles, mainly of quartz and igneous rocks. The tops of siltstone sequences are also erosional and channelled. The contact with overlying sandstones is marked by a thin conglomerate of both extraformational and intraformational red shale pebbles (Fig. 2.9).

Palaeocurrents determined from cross-bedding in the red facies on Raasay indicate deposition by southeasterly directed currents. There is a very low scatter of readings. Regionally, however, red facies palaeocurrents radiate from two point sources west of the present outcrop (Fig. 2.3). Deposits of the northern fan wedge out southwards beneath the second.

## Red facies: interpretation

The coarse grain size and cross-bedding of this facies indicate deposition by unidirectional high velocity traction currents. The lenticular geometry of the sand beds suggests sedimentation by alternate scouring and infilling of hollows. Shale laminae between sand lenses indicate that current velocity fluctuated widely. The shale lenses between erosion surfaces are clearly abandoned channel deposits. Considered overall therefore the evidence suggests that the red facies originated in an alluvial environment; the sandstones having been deposited from migrating megaripples by violent currents in active channels; and the shales originating in abandoned channels. Nearly half the red facies sandstones were penecontemporaneously deformed. Foresets are overturned downcurrent, original flat-bedding convoluted, whole sandstone beds form diapiric intrusions metres high, and heavy mineral bands founder down into arkose sand. The dominantly vertical trend of these structures suggests that they are not slumps, due to lateral movement, but fossilized quicksands due to water escaping from waterlogged sediment. The expulsion of pore water could have been initiated by seismic vibrations, by turbulent currents, or hydrostatic pressure generated by connate water migrating down the palaeoslope. All these processes have successfully generated analogous quicksand structures in the laboratory. The weight of the evidence suggests however that turbulent currents were the most probable initiator of quicksand movement. Deformational structures similar to those of the red facies (though generally smaller and less abundant) are ubiquitous in sands deposited by traction currents in many environments (Allen and Banks, 1972).

Considering the red facies as a whole, coarse grain size, low palaeocurrent variability, and absence of shale sheets with transitional bases comparable to overbank sediment all suggest deposition by low sinuosity braided channel complexes, rather than by high sinuosity meandering rivers. This explanation is consistent with the regionally radiating palaeocurrents which strongly suggest that the red facies was deposited on piedmont fans. It is interesting to note that the fan apices lie along the line of the Minch fault (evidence for the existence of which is based on a deep, glacially scoured, now subsea, valley and an offset of structural zones between the Lewisian gneiss of the Outer Hebrides and the Scottish mainland (Dearnley, 1962). This structure has been interpreted as a Caledonian tear fault analogous to that of the Great Glen. It is interesting to speculate, however, that the Minch fault had an easterly downthrow in the late PreCambrian. The Torridon Group piedmont fans may have been generated by its escarpment.

Geopoetry aside, it can be seen that these sediments are a good example of ancient alluvium attributable to deposition from braided rivers. Many other examples are known from around the world (e.g. Miall, 1981; Nilson, 1982).

## DEVONIAN SEDIMENTS OF SOUTH WALES AND THE CATSKILL MOUNTAINS, USA: DESCRIPTION

At the end of the Silurian Period the Caledonian orogeny formed extensive mountain chains in the continental areas around the North Atlantic. Marginal basins, often fault-bounded, were infilled during the Devonian Period by thick sequences of red conglomerates, sandstones, and shales (the Old Red Sandstone). These thin away from the mountainous source regions and pass into fine-grained marine sediments. Red-bed sedimentation was generally terminated by a marine transgression in the Lower Carboniferous. Devonian red beds have been intensively studied in North America, Greenland, Spitzbergen, and the United Kingdom (Fig. 2.10).

*Figure 2.10*   Photograph of Continental Devonian 'Old Red Sandstone'. Kong Oscars Fjord. East Greenland.

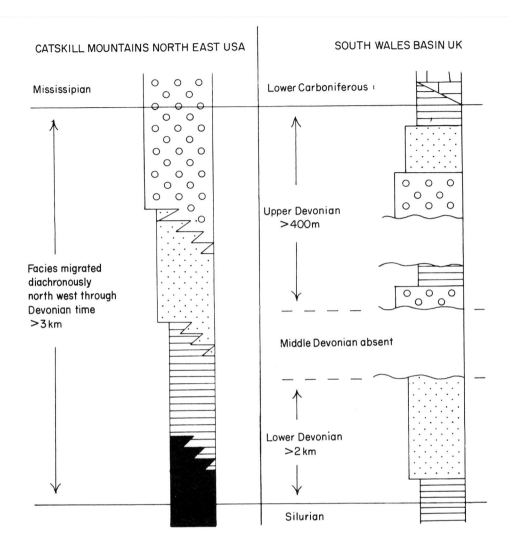

CATSKILL MOUNTAINS NORTH EAST USA          SOUTH WALES BASIN UK

Mississipian

Lower Carboniferous

Upper Devonian
>400m

Facies migrated
diachronously
north west through
Devonian time
>3km

Middle Devonian absent

Lower Devonian
>2km

Silurian

| Key | Catskills | S. Wales | Environment |
|---|---|---|---|
| | Pocono | Facies A | Braided Rivers |
| | Catskill | Facies B | Meandering Rivers |
| | Chemung | Facies C | Shoreline |
| | Portage | Absent | Open Marine |

*Figure 2.11* Generalized comparative sections of Catskill (USA), and south Wales Devonian sediments.

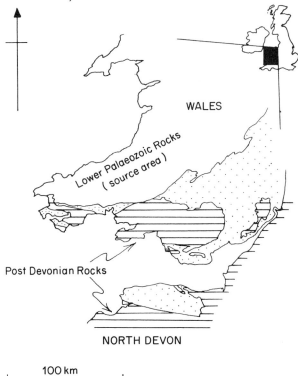

WALES

Lower Palaeozoic Rocks
( source area )

Post Devonian Rocks

NORTH DEVON

|_____ 100 km _____|

*Figure 2.12* Map of south Wales and adjacent areas showing outcrop of Devonian strata (stippled). Almost totally continental (Old Red Sandstone) facies in south Wales pass south-wards into interbedded marine and continental facies in north Devon.

*Table 2.1. Summary of Devonian facies in south Wales and Pennsylvania and New York State, USA.*

| Nomenclature | | | | |
|---|---|---|---|---|
| N.E. USA | S. Wales | Sedimentology | Fauna | Environment |
| Pocono | facies A | coarse red pebbly cross-bedded sand-stone, rare shales | generally barren | braided alluvium |
| Catskill | facies B | interbedded red medium sands and shales | fish, lamellibranchs plants | meandering alluvium |
| Chemung | facies C | interbedded grey fine sands and shales, rare coals | plant debris, burrows, lamellibranchs, and brachiopods | marine shoreline |
| Portage | — | laminated dark pyritic shales, rare dark calci-lutites and greywackes | ammonoids lamellibranchs | open marine |

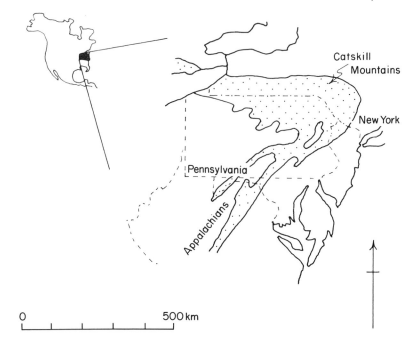

*Figure 2.13* Outcrop of Devonian sediments, northeastern USA.

Based on data from northeastern USA and south Wales four main Devonian facies can be recognized (Table 2.1). Their stratigraphy is summarized in Fig. 2.11, and their outcrops in Figs 2.12 and 2.13.

The Catskill facies (facies B of south Wales) will now be described. The reasons why it is thought to be the alluvium of meandering rivers will then be discussed.

In its type area of the Catskill Mountains this facies has a north-westerly thinning prismatic geometry with a maximum thickness of about 600 m. It is overlain by the westerly diachronous Pocono facies and itself diachronously overlies the Chemung facies. The analogous facies 'B' deposits of south Wales show a more erratic and less well-defined distribution. They occur at two stratigraphic levels being limited vertically both by unconformities and changes into Pocono- and Chemung-type sediments (Fig. 2.11).

Lithologically the Catskill facies consists of sandstones and shales in about equal proportions with a minor amount of conglomerates.

The conglomerates are thin, lenticular, and seldom more than a few decimetres thick. They are composed of extraformational pebbles of vein quartz and rock fragments, and of intraformational shale pebble of local penecontemporaneous origin.

The sandstones are variable in texture and composition. Generally they are medium- to fine-grained and seldom well sorted. A sparse clay matrix is

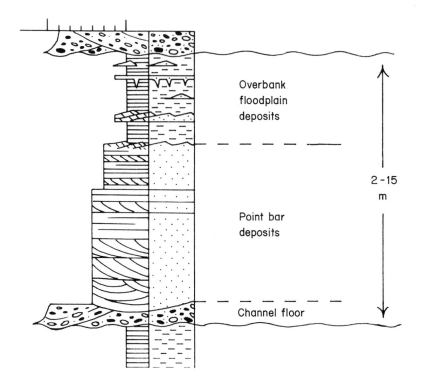

*Figure 2.14* Generalized section of fining-upward sequence in Devonian Catskill facies, USA and south Wales.

sometimes present. Petrographically these vary from lithic sandstones to sub-greywackes. In south Wales they are red-coloured due to ferric iron in the matrix. In the Catskill region they are generally drab or mottled.

The shales with which the sandstones are interbedded are argillaceous sandy siltstones with thin fine sand layers. A red colour predominates though drab and mottled units do occur. Interbedded with the shales are occasional thin layers of nodular microcrystalline limestone and dolomite. Pebbles of this lithology sometimes occur in the conglomerates due to reworking.

The three lithologies of the Catskill facies have a tendency to be repeatedly arranged in upward-fining sequences of conglomerate, sand, and shale. These lithologies are accompanied by a corresponding systematic vertical arrangement of sedimentary structures (Fig. 2.14).

Each sequence commences with an erosional, scoured, and often channelled surface cut into the shales beneath. This is overlain by the thin conglomerate member, though it may locally be absent. The conglomerate is generally massive or sometimes poorly stratified with bedding surfaces arranged sub-parallel to channel margins.

The sandstones above the conglomerates are generally medium-grained and cross-bedded in their lower part grading up into fine flat-bedded and micro-crosslaminated sandstone. The geometry of the sand bodies is generally hard to establish due to poor exposure. Sometimes however it is possible to see that they are either isolated channels or laterally extensive sheets composed of a plexus of laterally coalesced channels. Occasionally major bedding surfaces can be seen that dip in towards the channel axes. Between the major bedding surfaces are cross-beds that dip down channel. This is therefore a Type I fabric as described in Chapter One. The sandstones pass up transitionally into the shales. These generally show lamination and contain thin sandstone layers, ripples, sand-filled desiccation cracks, and nodular carbonate horizons. These fining-upward sequences of conglomerate sand and shale are generally between 2–15 m thick.

Dip directions of cross-bedding are highly variable at individual sample points. Plotted regionally, however, they generally show a preferred trend which is believed to correspond to the palaeoslope. This was southwards in the Welsh basin and generally northwestwards in the Catskills.

Fossils in this facies are generally hard to find. Shales sometimes contain plant fragments and rootlet horizons. The sandstones occasionally yield disarticulated plates and spines of primitive fish such as the Pteraspids and lamellibranch shells often, in south Wales, called *Archanodon jukesi*. Plant fragments are sometimes present in the sandstones and conglomerates.

## DEVONIAN SEDIMENTS OF SOUTH WALES AND THE CATSKILL MOUNTAINS, USA: INTERPRETATION

The repeated upward-fining of grain size in this facies suggests deposition by currents which at any one point waxed quickly and waned slowly in velocity through time. Thus the conglomerate covered erosion surfaces record scouring turbulent currents while the overlying sands indicate lower velocity unidirectional traction currents which deposited cross-beds from migrating megaripples. The gradation up into siltstone shows that velocity diminished to the extent that fine sediment could settle out of suspension. The channelled nature of the sands and the way in which major bedding surfaces dip towards the channel axis, invites comparison with the point bars of modern meander channels.

The interbedded shales with their cross-laminated fine sands and desiccation cracks imply sedimentation in an environment in which deposition from suspension and low velocity traction currents alternated with periods of subaerial exposure. The nodular cryptocrystalline carbonates are analogous to modern caliche soil horizons. Considered altogether it seems most probable that the fine grained subfacies was deposited in an overbank flood plain setting, through which the fluvial channels meandered (refer back to Fig. 2.1).

The fauna is generally believed to be freshwater. The reason for this is largely a circular one: that the deposits are continental, therefore the fossils are non-marine. It is significant, however, that there is an absence of undoubtedly marine fossils, such as brachiopods, echinoderms, and trilobites. Similarly the occurrence of plant debris and rootlet horizons suggests that the environment, if not continental, was certainly near the shore and intermittently swampy.

Considered altogether the evidence points to the conclusion that the Catskill deposits were laid down in a continental environment. The vertical sequence of lithologies and sedimentary structures is closely comparable to those produced today by the lateral migration of meandering rivers (p. 43); the conglomeratic scoured surfaces originating on channel floors, the sands being deposited on the point bars and the silts on the floodplains (Mackin, 1937; Allen, 1964; Visher, 1965).

This type of fining-upward cycle occurs in many ancient alluvial deposits (see Allen, 1965, for a review). Before this simple explanation was advanced several other suggestions had been made. These included eustatic changes of sea level, which would affect the base level and equilibrium of a river; tectonic uplift of the source, which would increase sediment supply; and climatic changes varying rainfall and hence runoff and current velocity.

It is clear, however, that the to and fro migration of a river across its floodplain is an adequate explanation for alluvial cycles when superimposed on gradual tectonic subsidence. This is not to say that the other processes did not affect sedimentation but that if they did then their effects must be superimposed upon this built-in cycle mechanism of an alluvial floodplain. Certainly the large-scale vertical facies variations and unconformities of the south Wales basin indicate forces beyond those generated by a meandering river. In a statistical study of Devonian cyclothems in Pembrokeshire, Allen (1974) showed that both localized and laterally extensive types were present. He attributed the former to autocyclic processes, such as channel migration, and the latter to allocyclic processes of regional significance.

Five upward-fining sequences occur in the Pleistocene and Recent alluvium of the Mississippi river valley. These can be attributed to eustatic sea-level changes during the ice age (Turnbull, Krinitsky, and Johnson, 1957). For further discussion of the origin of cycles in alluvial successions see Beerbower (1962), Allen (1964; 1965), Visher, (1965), and Duff, Hallam, and Walton (1967, pp. 21–35).

## DISCUSSION

Large volumes of rock in many parts of the world ranging in age from Pre-Cambrian to the present day have been attributed to alluvial environments. Basically these can be classified into four main types:

(1) Prisms of sediment thousands of feet thick deposited in basins adjacent to mountain chains.

(2) Sequences, thousands of feet thick, deposited in fault-bounded troughs within continental shields.

(3) Laterally extensive sheets of coarse braided alluvium generally only a few hundred feet thick blanketing continental shields.

(4) Thin sheets of alluvium beneath transgressive marine deposits.

The first type of alluvial deposit is shown by the Catskill and associated facies of North America and by the south Wales Old Red Sandstone basin just described. Other examples include the Upper Cretaceous sediments east of the Rocky Mountains (Chapter 6), the Tertiary molasse of Switzerland, and the Siwalik Series south of the Himalayas. These examples have the following features in common. They all were formed in basins along the edges of rising mountain chains. They are thousands of feet thick adjacent to the mountains from which they were derived. The sediments thin away from the source as grain size diminishes. Facies analysis suggests that the depositional environment changed from braided alluvium at the source, through meandering alluvium to marine shoreline and, ultimately, open marine conditions.

The second type of alluvium occurs in fault-bounded troughs often closed from the sea either within mountain chains or on continental shields. These deposits, often thousands of feet thick, are generally coarse braided alluvium generated from fans on the marginal fault scarps. They may pass laterally into meandering alluvium or directly into lacustrine deposits in the trough centre. These deposits are often interbedded with volcanic rocks which erupted along the marginal faults. Example of this model include the Triassic Newark: Connecticut trough complex of northeast North America (Klein, 1962; Van Houten, 1964), the Devonian Midland Valley of Scotland, the PreCambrian Dala and Jotnian sandstones of Scandinavia, and possibly, the Torridonian.

The third type of alluvial occurrence are sheets of coarse pebbly sandstone which, though only a few hundred feet thick, cover hundreds of square miles. At their base they are separated by a conglomerate from a planar unconformity though in detail this may be channelled or with protruding hillocks.

Extensive deposits of this type occur around the edges of the Saharan and Arabian Shield. These range in age from Cambrian to Recent (Picard, 1943), and are loosely referred to as of Nubian facies (the type Nubian is Cretaceous). Sedimentologically these rocks are closely comparable to the braided alluvium of the Torridonian red facies. It is difficult to see, however, how the steep gradients and high current velocity necessary for braided rivers could have been maintained downcurrent over hundreds of miles. One would expect sediment to quickly build up to base level so that current velocity diminished and fine-grained meandering alluvium deposited.

An answer to this problem was suggested by Stokes (1950) in a study of the Shinarump and similar formations on the Colorado plateau. Here there are several sheets of coarse sandstones and conglomerates each generally less than 300 ft thick and overlying a planar unconformity. Stokes points out that, since these rocks contain terrestrial fossils, it is unlikely that they were laid down by a sea transgressing over a peneplain. It is more probable that these sand sheets were deposited on piedmont fans derived from scarps which retreated as they cut back across a pediplain. Williams (1969) put forward a similar explanation for the Torridonian fans.

This concept can be applied to the Nubian facies of the Saharan and Arabian Shields. Typical examples of these deposits occur in the Southern Desert of Jordan (Fig. 2.15). Here the PreCambrian igneous basement is overlain by 700 m of coarse cross-bedded pebbly sandstones attributable to a braided alluvial environment (Selley, 1970; 1972). The alluvial facies is divisible into three formations (Fig. 2.16). The unconformity with the PreCambrian basement is a planar (?penecontemporaneously) weathered surface south of Wadi Rum. Downcurrent to the north inselbergs protrude 35 m into the overlying Saleb Formation which thickens northwards from 30 m south of Wadi Rum to about 60 m in a distance of 30 km. The base of the overlying Ishrin Formation is marked by huge channel complexes whose floors are lined with imbricate siltstone slabs over 1 m long. These were presumably re-worked from the Saleb Formation beneath. The Ishrin Formation is about 300 m thick and shows little regional thinning over hundreds of square miles. The top of the Ishrin Formation is incised by channels over 5 m deep which again are sometimes completely infilled by siltstone slabs. These channels mark the base of the Disi Formation which, like the Ishrin Formation, is about 300 m thick and shows little regional thickness variation over hundreds of square miles. At its top the Disi passes abruptly, but apparently without erosion, into overlying marine shelf sands of the Um Sahm Formation (p. 199).

*Figure 2.15* Pediment cut in Pre-Cambrian igneous basement (dark rock), overlain by Cambro-Ordovician braided alluvium of the Saleb and Ishrin formations. Recent braided alluvial fan in foreground. Wadi Rum, Jordan.

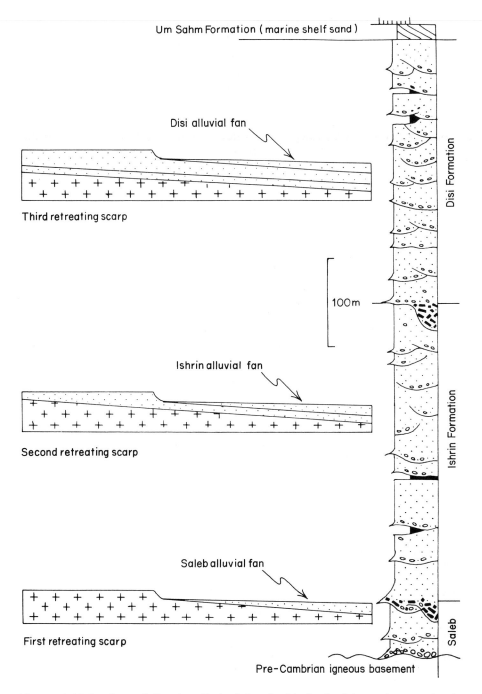

Um Sahm Formation (marine shelf sand)

Disi alluvial fan

Third retreating scarp

100m

Ishrin alluvial fan

Second retreating scarp

Saleb alluvial fan

First retreating scarp

Disi Formation

Ishrin Formation

Saleb

Pre-Cambrian igneous basement

*Figure 2.16* Section of Cambro-Ordovician braided alluvial sandstones in the Southern Desert, Jordan. Diagrams illustrate possible depositional mechanism of repeated uplift and pedimentation.

Deposition from the braided fans of repeatedly retreating scarps is certainly a most attractive explanation for these extensive alluvial deposits. According to this concept the Arabian Shield was uplifted three times. Each phase of rejuvenation of the landscape caused scarps to retreat into the hinterland of the shield. The first scarp would have cut a pediment into the basement on which the braided alluvium of the Saleb Formation was deposited. The next phase of uplift caused a second scarp to cut a new pediment into the Saleb deposits and bury it with Ishrin alluvium. A third repetition of this process deposited the Disi Formation (Fig. 2.16).

This process can explain the Nubian-type sand sheets of the Arabian and Saharan shields.

The fourth type of alluvial deposit, thin sheets beneath marine transgressions, is genetically related to shorelines and discussed therefore in Chapter 6.

## ECONOMIC ASPECTS

Alluvial deposits are of economic interest in the search for petroleum, uranium, metals, and coal.

Alluvial deposits can make good petroleum reservoirs where there are adjacent source beds. Since these may be absent, especially in arid continental basins, an unconformity generally separates petroliferous alluvial reservoirs from their

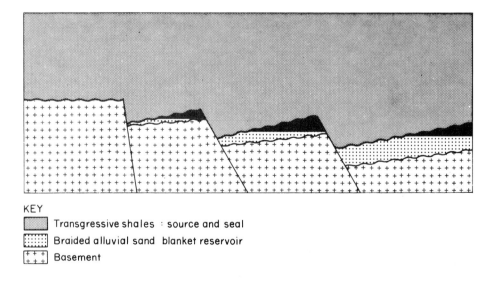

KEY

▨ Transgressive shales : source and seal
▦ Braided alluvial sand  blanket reservoir
⊞ Basement

*Figure 2.17* Cartoon to show how braided alluvial sand sheets may form major petroleum reservoirs in tilted fault blocks when transgressed by organic rich shales. Examples cited in text. Oil shown in black.

source rock. A major distinction must be made between the petroleum potential of the sand blankets of braided channel systems and the isolated channel sands of meander belts. The former give rise to giant structurally trapped fields, while the latter host small stratigraphically trapped fields.

Braided channel systems can deposit thick blankets of porous permeable sands with few impermeable shale beds. Subsequently the region may subside, often accompanied by faulting, and be drowned by a marine transgression. This may result in the deposition of organic rich muds. After burial petroleum may migrate from the shales and become trapped in truncated braided alluvial sands on the crests of anticlines or tilted fault blocks (Fig. 2.17). This sequence of events

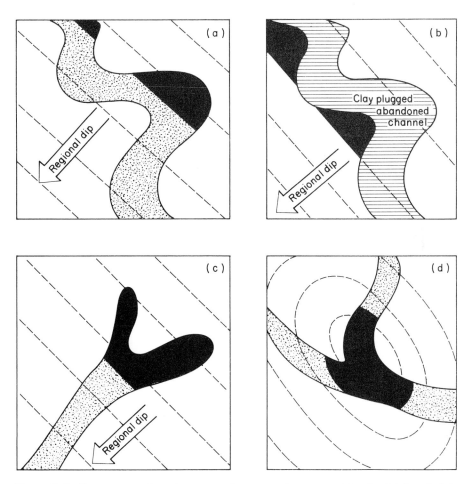

_Figure 2.18_ Cartoons to show how petroleum may be trapped in alluvial flood plain deposits. Note that though in most cases the reservoir is provided by sand within the channel, in case B the channel is clay-plugged and acts as a seal. Examples of the various types of trap are cited in the text.

has led to the development of such giant fields as Prudhoe Bay, Alaska (Jamison, Brockett and McIntosh, 1980), Hassi Messaoud, Algeria (Balducci and Pommier, 1970), and the Sarir and Messla fields of Libya (Sanford, 1970, Clifford *et al.*, 1980).

Because of the high ratio of shale to sand the alluvium of ancient meander channel flood plain systems can seldom host giant structural petroleum traps. Smaller stratigraphic accumulations are the norm. The limits of such fields may conform solely to the boundaries of a sand-filled channel. Usually however there is an element of closure due to regional tilt. Figure 2.18 illustrates four common types of trap in alluvial flood plain deposits.

Sometimes a sand-filled meander channel has been tilted in the direction of the palaeostrike. Petroleum migrating up dip may then be trapped in the updip termination of point-bar sands where they are sealed laterally by overbank shales (Fig. 2.18A). The Recluse field of Wyoming may be an example of this type of trap (Forgotson and Stark, 1972).

It is important to remember that a channel is an environment for transporting sand, but is not always an environment in which sand is deposited. Some channels become abandoned and clay plugged. Thus there are cases where abandoned channel shales have provided an updip seal to petroleum migration (Fig. 2.18B). Examples of this type of trap include the Coyote Creek and Miller Creek fields of Wyoming (Berg, 1968).

If tilting is in the direction of the palaeodip then oil may migrate far up a sand-filled channel and become trapped where it finally pinches out between impermeable beds above and below (Fig. 2.18C). The Clareton and Fiddler Creek fields of Wyoming are of this type.

Lastly channels may cross-cut anticlines giving rise to very elusive accumulations (Fig. 2.18D). The Pikes Peak field of Saskatchewan illustrates this type of trap (Putnam, 1983).

Aside from oil and gas, alluvial deposits can also be metalliferous. The gold deposits of Witwatersrand in South Africa are a case in point. These have been extensively described and discussed in the literature (e.g. Haughton, 1964; Pretorius, 1979). The Rand basin lies on the PreCambrian basement of South Africa. It measures about 250 km from northeast to southwest and 170 km from northwest to southeast. It is infilled by over 8 km of PreCambrian clastics which coarsen upwards and northwestwards towards their presumed source. There has been continuous debate about the depositional environment of the Witwatersrand deposits and of the genesis of the gold and uranium that they contain. Mapping of pebble size and palaeocurrent directions has shown that the sediments were deposited in a series of fans that radiated out into the basin from several points along the western margin (Fig. 2.19). Current opinion favours a braided alluvial outwash plain depositional environment. Regardless of whether the ore is a detrital placer or of syngenetic origin there is no doubt about the close correlation between its distribution and sedimentary facies. Regionally the ore is richest near fan heads, but on a local scale is concentrated in conglomerate channels some 600 m wide and 60 m deep.

Uranium mineralization occurs in Triassic-Jurassic alluvium in the Colorado Plateau and in Eocene alluvium in Wyoming, USA. Again there is debate as to the origin of the ores, but widespread agreement about the close correlation between mineralization and sedimentary facies. On a regional scale ore bodies occur in arcuate zones, halfway down ancient alluvial fans. There appears to have been a critical ratio of permeable sands to impermeable shales which favoured mineral precipitation (Crawley, 1982). This has been noted for example in the Uraven mineral belt within the Morrison alluvial fan of Colorado (Fischer, 1974), and in the Puddle Springs fan of Wyoming (Galloway, 1979). On a local scale the ore occurs in meniscus shaped bodies termed 'roll fronts' within fluvial

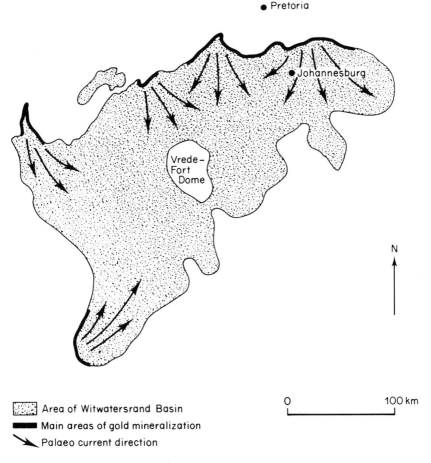

*Figure 2.19* Map of the Witwatersrand basin (Precambrian), South Africa, showing the correlation between gold and alluvial fan heads (based on Haughton, 1964 and Pretorius, 1979).

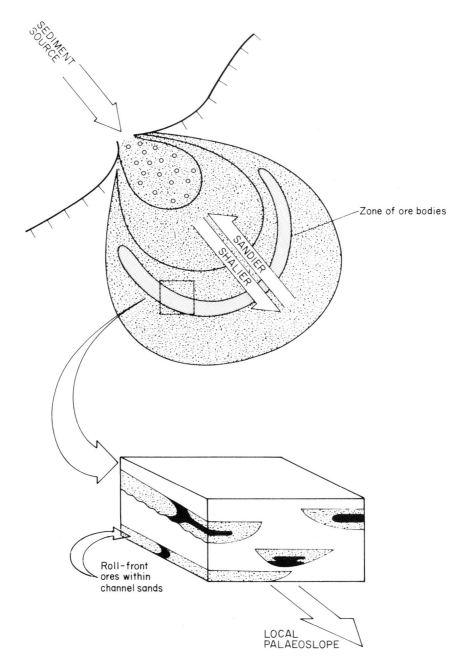

SEDIMENT SOURCE

SANDIER

SHALIER

Zone of ore bodies

Roll-front
ores within
channel sands

LOCAL
PALAEOSLOPE

*Figure 2.20* Cartoons to illustrate the sedimentological control on the distribution of uranium in alluvial fans. For examples see text.

channels. Precipitation of the ore probably occurred at the interface of mixing connate and uranium-enriched meteoric waters (Fig. 2.20).

Coal occurs in ancient alluvial deposits. For convenience sake the application of environmental interpretation to coal will be dealt with in Chapter 5, which deals with deltas.

## SUB-SURFACE DIAGNOSIS OF FLUVIAL DEPOSITS

Fluvial deposits are frequently encountered in the sub-surface, and since, as just discussed, they are sometimes prospective, it is important to be able to recognize them.

Since the deposits of braided and meandering channel systems are so different it is appropriate to consider them separately, reviewing the five parameters of geometry, lithology, sedimentary structures, palaeocurrents (i.e. dip motifs) and palaeontology for each in turn.

### Sub-surface diagnosis of braided fluvial deposits

The geometry of braided fluvial deposits tends to be sheet-like, thick, laterally extensive and often overlying an irregular or pedimented unconformity surface.

The upper and lower boundaries of braided alluvial deposits may show up as seismic reflectors where they generate sufficient velocity contrast with the adjacent rocks. Blanket or fan geometries may be mappable. Because of the uniform sandy nature of braided alluvial deposits they contain few internal velocity variations and thus seldom show any internal reflectors.

The lithology of braided fluvial deposits consists almost entirely of conglomerates and coarse sands, with only minor amounts (some ten per cent) of fine sands and siltstones. Glauconite is absent naturally, but so also is carbonaceous organic matter due to the oxidizing nature of the environment. Braided alluvial sediments tend to be red in colour due to the presence of a red ferric oxide cement. The origin of red colouring in sediments is due to

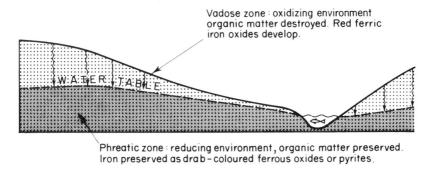

Vadose zone : oxidizing environment organic matter destroyed. Red ferric iron oxides develop.

W·A·T·E·R··T·A·B·L·E

Phreatic zone : reducing environment, organic matter preserved. Iron preserved as drab-coloured ferrous oxides or pyrites.

*Figure 2.21* Cartoon to show the relationship between early diagenesis of iron compounds and the level of the water table.

diagenesis, which is beyond the scope of this book (see Turner, 1980). As a general statement however sediments that are deposited above the water table undergo early diagenesis in an oxidizing environment. Organic matter is destroyed and iron preserved as red ferric oxide. Sediments that are deposited below the water table undergo early diagenesis in a reducing environment. Organic matter may be preserved and iron preserved as drab-coloured ferrous oxides or pyrites (Fig. 2.21).

Thus braided alluvial fan sands tend to be red coloured, as do some other deposits of arid or semiarid continental environments. There are exceptions to these general statements (naturally). Some deep marine oozes are red (p. 284) and secondary reddening occurs beneath some unconformities. Red beds may become grey-green in colour if they are flushed by strongly reducing connate fluids. This phenomenon is especially common adjacent to petroleum accumulations.

The characteristic sedimentary structure of braided alluvial deposits is channelling, though this cannot be seen in a single well log or core. The channels

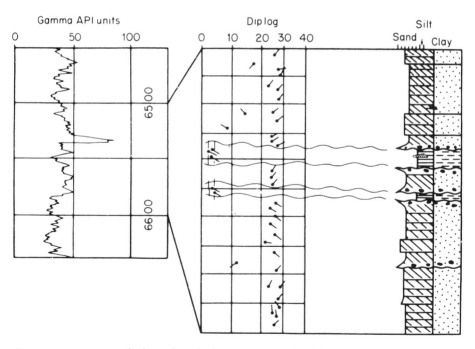

*Figure 2.22* Log motifs for a borehole penetrating braided alluvial deposits. The gamma log shows that this is a more or less uniform sandy sequence with a single shale unit. The dip log shows green structural dips in the shales and steep foreset dips in the channel sands. Core logs show mainly cross-bedded sands with an abandoned channel shale lying between two erosion surfaces.

are infilled with cross-bedded and flat-bedded sands, with occasional recumbent foresets and disturbed bedding.

The most characteristic feature to look for in cores is the double erosion surface above and below shale units. These abandoned channel sequences are diagnostic of braided deposits. Recognition of these is important because one may predict that the shales have shoestring geometries and that they will not act as barriers to fluid migration in hydrocarbon reservoirs or aquifers.

Dipmeter motifs of braided deposits tend to be complex. Green patterns indicative of structural dip may be present in the abandoned channel shales. 'Bag o' nails' motifs are not uncommon in the channel conglomerates and sands, but 20–25° foreset beds may be present which dip in the direction of current flow.

Cross-bedding in braided channel systems tends to show a well-developed down current vector mode (Smith, 1972).

Quite large changes in mean direction of dip are common from channel to channel.

Braided alluvial deposits tend to be palaeontologically barren because of the oxidizing nature of the environment. There may be rare vertebrate tracks and trails, including orgasmoglyphs produced by rutting dinosaurs.

These then are the five criteria to look for in the sub-surface identification of braided deposits. Figure 2.22 shows the gamma profile, dip log pattern and sedimentary core log characteristics to be anticipated in a borehole penetrating such a facies.

## Sub-surface diagnosis of meandering fluvial deposits
In the same way that braided channels pass into meandering ones, so their deposits show a similar gradation. True meander flood plain sediments differ from braided outwash plains principally in their sand:shale ratio. Whereas

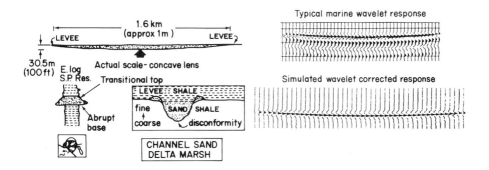

*Figure 2.23* Geological model and seismic response for an upward-fining channel (from Schramm *et al.* 1977 by courtesy of the American Association of Petroleum Geologists).

braided outwash plains are almost exclusively composed of sand and gravel, the alluvium of meander channel flood plains consist of about equal parts channel sand and flood plain shale.

The external geometry of the two types may be comparable, but the internal characters are very different. The channelled nature of flood plain sediments may generate a series of impersistant seismic reflectors. Since channels commonly have an erosional base and a gradational top detailed analysis suggests that these signals are generated by the channel floors (Fig. 2.23). Three-dimensional

*Figure 2.24* Three-dimensional seismic survey from the Gulf of Thailand, showing alluvial flood plain meander belt (from Brown *et al.*, 1982, by courtesy of the American Association of Petroleum Geologists).

seismic surveys may be able to delineate a whole meander belt in which channels and ox-bow lakes can be differentiated from the overbank shales (Fig. 2.24).

Lithologically meandering fluvial deposits have an overall finer grain size, with sand: shale ratios of about 50:50. Conglomerates are rare, apart from those of intraformational origin. The sands tend to be relatively finegrained. Red colouration may be present in semi-arid alluvial deposits, and may be associated with nodular carbonate caliches. Drab-coloured sand and shales, reflecting waterlogged alluvial plains, occur however, and these are often associated with coal beds, and disseminated carbonaceous detritus in sands and shales alike.

Cored intervals should show the suites of sedimentary structures associated with upward-fining grain-size sequences described previously. A channel floor erosion surface may be followed by cross-bedded point-bar sands which grade into crosslaminated finer sands and desiccation-cracked overbank shales.

Dipmeter patterns are of two types. Green motifs indicative of structural dip occur in the shales, while upward-decreasing red motifs characterize channels sands. Detailed analysis of the red patterns sometimes shows a bimodal pattern: one mode pointing towards the channel axis due to major point-bar bedding

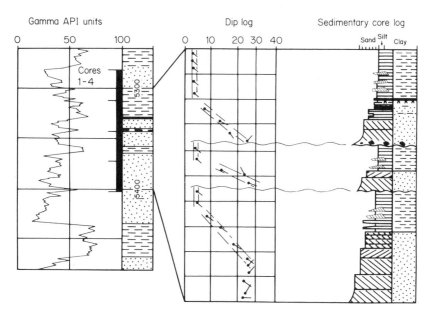

*Figure 2.25* Log motifs for a borehole penetrating alluvium of meandering fluvial systems. Regular upward-fining motifs are not readily apparent on the gamma log, but the high ratio of shale to sand differentiates these deposits from continental beds of braided channel or eolian origin. Note the green dip motifs in the shales and the red patterns in the channel sands. Upward-fining grain-size profiles are readily seen in the core.

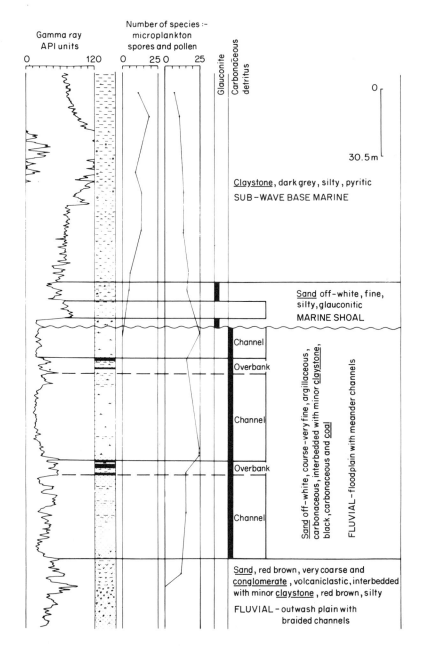

Gamma ray
API units

0        120

Number of species :-
microplankton
spores and pollen

0        25 0        25

Glauconite

Carbonaceous detritus

0

30.5 m

Claystone, dark grey, silty, pyritic
SUB-WAVE BASE MARINE

Sand off-white, fine,
silty, glauconitic
MARINE SHOAL

Channel

Overbank

Channel

Overbank

Channel

Sand off-white, course-very fine, argillaceous, carbonaceous, interbedded with minor claystone, black, carbonaceous and coal

FLUVIAL - floodplain with meander channels

Sand, red brown, very coarse and
conglomerate, volcaniclastic, interbedded
with minor claystone, red brown, silty

FLUVIAL - outwash plain with
braided channels

*Figure 2.26* Borehole from the North Sea showing fluvial deposits unconformably overlain by marine shales. There are several cycles in which channel sands with abrupt erosional bases pass up with a steady or upward shaling gamma log motif into overbank shales and coals. From Selley (1976, Fig. 5) by courtesy of the *American Association of Petroleum Geologists*.

surfaces, the second mode, at 90° to the first, pointing down channel, reflecting cross-beds. This dip motif is illustrated and discussed in more detail in the context of deltaic distributaries with which they are analogous (p. 142).

Palaeontologically meandering deposits may yield non-marine pollen and spores.

Figure 2.25 shows the log motifs which may be anticipated in boreholes which penetrate the deposits of meandering fluvial systems and Fig. 2.26 illustrates a well log from the North Sea that penetrated several increments of channel sands and overbank coals and shales.

## REFERENCES

The description of the Torridonian sediments was based on:

Dearnley, R., 1962. An outline of the Lewisian complex of the Outer Hebrides in relation to that of the Scottish mainland. *Quart. J. Geol. Soc. Lond.*, **118**, 143–66.
Parkes, L. R., 1963. The Origin of the Moon. *New Scientist*, **20**, No. 365, p. 404.
Selley, R. C., 1964. The Penecontemporaneous deformation of heavy mineral bands in the Torridonian Sandstone of N.W. Scotland. In: *Deltaic and Shallow Marine Sediments.* (Ed. L. M. J. U. Van Straaten) Elsevier, Amsterdam, 362–7.
——, 1965a. Diagnostic characters of fluviatile sediments of the Torridonian Formation (PreCambrian) of North-West Scotland. *J. Sediment. Petrol.*, **35**, 366–80.
——, 1965b. The Torridonian Succession on the Islands of Fladday, Raasay and Scalpay, Inverness-shire. *Geol. Mag.*, **102**, 361–9.
——, 1966. Petrography of the Torridonian rocks of Raasay and Scalpay, Inverness-shire. *Proc. Geol. Ass.*, **77**, 293–314.
——, 1969. Torridonian Alluvium and Quicksands. *Scott. J. Geol.*, **5**, No. 4, 328–46.
——, and Shearman, D. J., 1962. Experimental Production of Quicksands. *Proc. Geol. Soc.*, *London*, No. 1599, 101–2.
——,——, Sutton, J., and Watson, J., 1963. Some Underwater Disturbances in the Torridonian of Skye and Raasay. *Geol. Mag.*, **100**, 224–43.
——, and Stewart, A. D., 1967. The Torridonian Sediment of N.W. Scotland. *Field Guide for Excursion C.2. 7th Internat. Sed. Congress.*
Stewart, A. D., 1966. An Unconformity in the Torridonian. *Geol. Mag.*, **103**, 462–5.
——, 1969. Torridonian rocks of Scotland. Reviewed in: North Atlantic geology and continental drift. (Ed. M. Kay) *Amer. Assoc. Petrol. Geol. Tulsa.* 1082 p. pp. 595–608.
——, 1975. 'Torridonian' rocks of western Scotland. *Spec. Rep. Geol. Soc. London.*, **6**, 43–51.
——, 1982. Late Proterozoic rifting in North West Scotland: the genesis of the Torridonian. *J. Geol. Soc. Lond.*, **139**, 413–20.
——, and Parker, A., 1979. Palaeosalinity and environmental interpretation of red beds from the Late Precambrian ('Torridonian') of Scotland. *Sedimentary Geology*, **22**, 229–41.
Van Houten, F. B., 1961. Climatic Significance of Red Beds. In: *Descriptive Palaeoclimatology.* (Ed. A. E. M. Nairn) Interscience, N.Y., 89–139.
Williams, G. E., 1969. Characteristics and Origin of a PreCambrian Pediment. *J. Geol.*, **77**, 183–207.

The account of the Devonian deposits of the Catskills, USA, and the Old Red Sandstone of south Wales was based on:

Allen, J. R. L., 1964. Studies in fluviatile sedimentation: six cyclothems from the Lower Old Red Sandstone, Anglo-Welsh basin. *Sedimentology*, **3**, 163–98.

——, 1965. Upper Old Red Sandstone Paleogeography in South Wales and the Welsh Borderland. *J. Sediment. Petrol.*, **35**, 167–96.

——, 1970. Studies in fluvial sedimentation: a comparison of fining-upwards cyclothems, with special reference to coarse member composition and interpretation. *J. Sediment Petrol.*, **40**, 298–323.

——, 1974. Studies in fluviatile sedimentation: lateral variation in some fining-upwards cyclothems from the Red Marls. *Geol. Jour.*, **9**, 1–16.

——, 1983. Studies in fluviatile sedimentation: bars, bar complexes and sandstone sheets (low sinuosity braided streams) in the Brownstones (L. Devonian), Welsh Borderlands. *Sedimentary geology*, **33**, 237–93.

——, and Crowley, S. F., 1983. Lower Old Red Sandstone fluvial depositional systems in the British Isles. *Trans. Roy. Soc. Edinb.*, **74**, 61–68.

——, and Friend, P. F., 1968. Deposition of the Catskill Facies, Appalachian Region: with notes on some other Old Red Sandstone Basins. In: Late Palaeozoic and Mesozoic Continental Sedimentation. (Ed. G. de V. Klein) *Geol. Soc. Amer. Sp. Pap.*, No. 206, 21–74.

——, and William, B. P. J., 1982. The architecture of an alluvial suite: rocks between the Townsend Tuff and Pickard Bay Tuff Beds (Early Devonian). S. W. Wales *Phil. Trans. Ser. B.*, **297**, 51–89.

Friedman, G. M., and Johnson, K. G., 1966. The Devonian Catskill Deltaic Complex of New York, type example of a 'tectonic delta complex'. In: *Deltas*. (Ed. M. L. Shirley and J. A. Ragsdale) Houston Geol. Soc., 171–88.

——, and ——, 1969. The Tully Clastic Correlatives (Upper Devonian) of New York State: A Model for Recognition of Alluvial Dune (?), Tidal, Nearshore (Bar and Lagoon), and Offshore Sedimentary Environments in a Tectonic Delta Complex. *J. Sediment. Petrol.*, **39**, 451–85.

Friend, P. F., 1966. Clay fractions and colours of some Devonian Red beds in the Catskill Mountains, U.S.A. *Quart. J. Geol. Soc. Lond.*, **122**, 273–92.

——, 1969. Tectonic features of Old Red Sedimentation in the North Atlantic Borders. In: North Atlantic-Geology and Continental Drift. (Ed. M. Kay) *Amer. Assoc. Petrol. Geol. Mem.*, **12**, 703–10.

——, and Moody, Stuart, M., 1972. Sedimentation of the Wood Bay formation (Devonian) of Spitsbergen: Regional analysis of a late orogenic basin. *Norsk Polarinstitutt*, Skrift 157, p. 77.

Humphreys, M. and Friedman, G. M., 1975. Late Devonian Catskill deltaic complex in north-central Pennsylvania. In: *Deltas: Models for Exploration*. (Ed. M. L. Broussard) *Houston Geol. Soc.* 53–84.

Other references quoted in this chapter were:

Allen, J. R. L., 1965a. Fining upward cycles in alluvial succession. *Lpool. Manchr. Geol. J.*, **4**, 229–46.

——, 1965b. A review of the origin and characteristics of Recent alluvial sediments. *Sedimentology*, **5**, No. 2, Sp. Issue. pp. 89–191.

——, and Banks, N. L., 1972. An interpretation and analysis of recumbent-folded deformed cross-bedding. *Sedimentology*, **19**, No. 3/4, 173–210.

Balducci, A. and Pommier, G., 1970. Cambrian Oil Field of Hassi Messaoud, Algeria. In: _Geology of Giant Petroleum Fields._ (Ed. M. T. Halbouty) _Amer. Assoc. Petrol. Geol. Mem._, **14**, 477–88.

Beerbower, J. R., 1964. Cyclothems and cyclic depositional mechanisms in alluvial plain sedimentation. In: Symposium on Cyclic Sedimentation. (Ed. D. F. Merriam) _Kansas Geol. Surv. Bull._, 169, **1**, 31–42.

Berg, W. A., 1968. Point bar origin of Fall River sandstone reservoirs, northeastern Wyoming. _Amer. Assoc. Petrol. Geol. Bull._, **52**, 2116–22.

Blackwelder, E., 1928. Mudflow as a Geologic Agent in Semi-arid Mountains. _Bull. Geol. Soc. Amer._, **39**, 465–83.

Blissenbach, E., 1954. Geology of Alluvial Fans in semi-arid Regions. _Bull Geol. Soc. Amer._, **65**, 175–90.

Brown, A. R., Graebner, R. J., and Dahm, C. G., 1982. Use of horizontal seismic sections to identify subtle traps. Mem. 32, _Amer. Assoc. Petrol. Geol._ (Ed. M. T. Halbouty) 47–56.

Cant, D. J., 1982. Fluvial Facies Models. In: _Sandstone Depositional Environments._ (Eds P. A. Scholle and D. Spearing) _Amer. Assoc. Petrol. Geol. Tulsa._, 115–38.

Chorley, R. J., (Ed.), 1969. _Introduction to Fluvial Processes._ Methuen, London, p. 218.

Clifford, H. J., Grund, R., and Musrati, H., 1980. Geology of a stratigraphic giant: Messla oil field, Libya. In: _Giant Oil and Gas Fields of the Decade 1968–78._ (Ed. M. T. Halbouty) _Amer. Assoc. Petrol. Geol. Mem._, **30**, 507–22.

Collinson, J. E. and Lewin, J., 1983. _Modern and Ancient Fluvial Systems. Sp. Pub. Internat. Assn. Sedol._ **6**, p. 584.

Costello, W. R. and Walker, R. G., Pleistocene sedimentology Credit river, Southern Ontario: a new component of the braided river model. _J. Sediment. Petrol._, **42**, 389–400.

Crawley, R. A., 1982. Sandstone Uranium Deposits in the U.S.A. _Energy Exploration and Exploitation_, **1**, 203–43.

Dahm, C. G., and Graebner, R. J., 1982. Field development with three-dimensional seismic methods in the Gulf of Thailand. _Geophysics_, **47**, 161–72.

Doeglas, D. J., 1962. The structure of sedimentary deposits of braided rivers. _Sedimentology_, **1**, 167–90.

Denny, C. S., 1967. Fans and Pediments. _Am. J. Sci._, **265**, 81–105.

Duff, P. McL. D., Hallam, A., and Walton, E. K., 1967. _Cyclic Sedimentation._ Elsevier, Amsterdam, p. 280.

Fischer, P. P., 1974. Exploration guides to new uranium districts and belts. _Econ. Geol._, **65**, 778–84.

Fisk, H. N., 1944. Geological investigation of the Alluvial Valley of the Lower Mississippi River. _Mississippi River Commission, Vicksburg._ p. 78.

Forgotson, J. M. and Stark, P. H., 1972. Well-data files and the computer, a case history from the northern Rocky Mountains. _Bull. Amer. Assoc. Petrol. Geol._, **56**, 1114–27.

Galloway, W. E., Kreitler, C. W., and McGowan, J. H., 1979. Depositional and ground water flow systems in the exploration for Uranium. _Bur. Econ. Ecol. Univ. Texas._, Austin Research Colloquium.

Gregory, K. J., (Ed.), 1977. _River Channel Changes._ Wiley, New York, p. 448.

Haughton, S. A. (Ed.), 1964. The geology of some Ore Deposits in Southern Africa. _Geol. Soc., South Africa_, 2 vols.

Jamison, H. C., Brockett, L. D. and McIntosh, R. A., 1980. Prudoe Bay: a ten year perspective. In: _Geology of Giant Petroleum Fields._ (Ed. M. T. Halbouty) _Amer. Assoc. Petrol. Geol. Mem._, **30**, 289–314.

Klein, G. de V., 1962. Triassic Sedimentation, Maritime Provinces, Canada. _Bull. Geol. Soc. Amer._, **73**, 1127–46.

Knight, M. J., 1975. Recent crevassing of the Erap River, Papau New Guinea. *Austral. Geogr. Studies*, **13**, 77–82.

Leopold, L. B., Wolman, M. G., and Miller, J. P., 1964. *Fluvial Processes in Geomorphology*. Freeman, San. Fran., p. 522.

Lowell, J. D., 1955. Applications of Cross-stratification studies to problems of Uranium Exploration. Chuska Mountains, Arizona. *Econ. Geol.*, **50**, 117–85.

Mackin, J. H., 1937. Erosional History of the Big Horn Basin, Wyoming. *Bull. Soc. Amer.*, **48**, 813–94.

Miall, A. D., 1977. A review of the braided river depositional environment. *Earth Sci. Rev.*, **13**, 62.

——, 1981. Sedimentation and tectonics in alluvial basins. *Geol. Soc. Can. Sp. Pap.*, **23**, 272.

Nilson, T. H., 1982. Alluvial fan deposits. In: *Sandstone Depositional Environments*. (Eds P. A. Scholle and D. Spearing). *Amer. Assoc. Petrol. Geol.*, Tulsa, 49–86.

Nininger, R. D. (Ed.), 1956. *Exploration for Nuclear Raw Materials*. Macmillan, London, p. 293.

Picard, L., 1943. *Structure and Evolution of Palestine*. Geol. Dept. Hebr. Univ., Jerusalem, p. 134.

Picard, M. D., and High, L. R., 1973. *Sedimentary structures of ephemeral streams*. Elsevier, Amsterdam, p. 239.

Pretorius, D. A. 1979. The depositional environment of the Witwatersrand Goldfields: A Chronological Review of Observations and Speculations. In: *Some Sedimentary Basins and Associated Ore deposits of South Africa. Geol. Soc. South Africa. Sp. Pub.*, **9**, 33–56.

Putnam, P. E., 1983. Fluvial deposits and hydrocarbon accumulations: examples from the Lloydminster area, Canada. In: *Modern and Ancient Fluvial Systems*. (Eds J. D. Collinson and J. Lewin) *Sp. Pub. No. 6, Internat. Assn. Sedol.*, 517–32.

Rachoki, A. H., 1981. *Alluvial Fans*. Wiley, Chichester, p. 172.

Rust, B. R., 1972. Structure and process in a braided river. *Sedimentology*, **18**, 221–46.

——, 1979. Facies Models 2 — Coarse alluvial deposits. In: *Facies Models*. (Ed. R. G. Walker). Geoscience Canada. Reprint Ser. 1. 9–21.

Sanford, R. M., 1970. Sarir oil field, Libya — Desert surprise. In: *Geology of Giant Petroleum Fields* (Ed. M. T. Halbouty) *Amer. Assoc. Petrol. Geol. Mem.* **14**, 449–76.

Schramm, M. W. Dedman, E. V. and Lindsey, J. P., 1977. Practical stratigraphic modelling and interpretation. In: *Seismic stratigraphy-applications to hydrocarbon exploration*. (Ed. C. E. Payton) *Amer. Assoc. Petrol. Geol. Mem.*, **26**, 477–502.

Schumm, S. A. 1977. *The Fluvial System*. Wiley, New York, p. 338.

Schumm, S. A., 1981. Evolution and Response of the fluvial system, sedimentological implications. In: *Recent and Ancient Non-marine environments: models for exploration*. (Eds F. G. Ethridge and R. M. Flores) *Soc. Econ. Pal. Min. Sp. Pub.*, **31**, 19–30.

Selley, R. C., 1970. Ichnology of Palaeozoic sandstones in the Southern Desert of Jordan: a study of trace fossils in their sedimentologic context. In: *Trace Fossils*. (Eds J. C. Harper and T. P. Crimes) Lpool. geol. Soc. 477–88.

——, 1972. Diagnosis of marine and non-marine environments from the Cambro-ordovician sandstones of Jordon. *Jl. geol. Soc. Lond.* **128**, 109–17.

——, 1976. Sub-surface environmental analysis of North Sea sediments. *Bull. Amer. Assoc. Petrol. Geol.*, **60**, 184–95.

Shanster, E. V., 1951. Alluvium of river plains in a temperate zone and its significance for understanding the laws governing the structure and formation of alluvial suites. *Akad. Nauk. S.S.S.R. Geol. Ser.* **135**, 1–271.

Smith, D. G., 1983. Anastomosed fluvial deposits: modern examples from western Canada. In: *Modern and Ancient Fluvial Systems.* (Eds J. D. Collinson and J. Lewin). *Sp. Pub. Internat. Assn. Sedol. No. 6,* 155–68.

Smith, N. D., 1972. Some sedimentological aspects of planar cross stratification in a sandy braided river. *J. Sediment. Petrol.,* **42,** 624–34.

Stokes, W. L., 1960. Pediment Concept Applied to Shinarump and similar Conglomerates. *Bull. Geol. Soc. Amer.,* **61,** 91–8.

Turnbull, W. J., Krinitsky, E. S., and Johnson, L. J., 1950. Sedimentary Geology of the Alluvial Valley of the Mississippi River and its bearing on foundation problems. In: *Applied Sedimentation.* (Ed. P. D. Trask) Wiley, N.Y., 210–26.

Turner, P., 1980. *Continental Red Beds.* Elsevier, Amsterdam, p. 562.

Van Houten, F. B., 1964. Cyclic Lacustrine Sedimentation, Upper Triassic Lockatong Formation. Central New Jersey and adjacent Pennsylvania. In: Symposium on Cyclic Sedimentation. (Ed. D. F. Merriam) *Geol. Surv. Kansas Bull.,* 169, **2,** 497–532.

Visher, G. S., 1965. Fluvial Processes as interpreted from ancient and recent fluviatile deposits. *Soc. Econ. Pal. Min. Sp. Pub.,* **12,** 116–32.

Williams, G. E., 1970. Piedmont sedimentation and Late Quarternary chronology in the Biska region of the northern Sahara. *Zeitschrift für Geomorphologie,* **10,** 40–63.

——, 1971. Flood deposits of the sandbed ephemeval streams of Central Australia. *Sedimentology,* **17,** 1–40.

Williams, P. F., and Rust, B. R., 1969. The Sedimentology of a braided river. *J. Sediment. Petrol.,* **39,** 649–79.

Woncik, J. 1972. Recluse field, Campbell County, Wyoming. In: *Stratigraphic Oil and Gas Fields.* (Ed. R. E. King) *Amer. Assoc. Petrol. Geol. Mem.,* **16,** 376–382.

# 3    Wind-blown sediments

## INTRODUCTION

Ancient wind-deposited sediments are often very hard to distinguish from those laid down by running water. As this chapter shows, however, a number of criteria have been proposed to distinguish eolian from aqueous sediments. Some of these are of debatable merit when used in isolation. The ultimate decision of whether a sediment is of eolian origin must be based on a critical evaluation of all the available data. Fortunately this is helped by studies of Recent wind-blown deposits, notably by Bagnold (1941), McKee (1966, 1977), Glennie (1970), Wilson (1972, 1973) and Ahlbrandt and Fryberger (1981, 1982).

When discussing sedimentary environments in an earlier section it was pointed out that most non-marine environments are areas of erosion. Most deserts are equilibrial surfaces. That is to say regions of the earth's crust where there is neither erosion, nor deposition taking place; though there may be a considerable amount of sediment being reworked and transported across desert surfaces. Unfortunately the Hollywood image of deserts as regions of sand dunes is a distortion of the facts, probably only about 25% of deserts are composed of sand dunes. The major surface areas are flat sand, gravel or bedrock covered plains.

### Recent eolian sediments

When the wind blows across a desert silt and clay are carried up into the atmosphere and may be transported far away ultimately to settle out in the seas. Any dust that does resettle in the desert may form beds of loess, as in periglacial realms, or be carried by floods into ephemeral desert lakes. Sand, by contrast, is transported close to the ground largely by saltation. Fortunately, wind velocities are seldom sufficiently high to transport gravel.

The bed forms of sands transportation are asymmetric ripples and megaripples whose geometry resembles that of subaqueous ones. The physics of eolian and aqueous transportation are basically similar concerning a granular solid within a fluid. This is probably the main reason why the eolian environment is the hardest one to recognize in the subsurface.

There are many different types of dune shape. Arbitrarily four main types can be defined though there are transitional varieties and some minor ones (Fig. 3.1).

Probably one of the best documented is the barchan or lunate dune. As with most dunes they have a gentle windward slope and a steep accretionary foreset on their downwind side. Barchans are characterized by two horns which are directed downwind at either end of the slipface. Barchans are very photogenic and photos are common in many text books. Their internal structures have also been studied in some detail. This effort is probably largely unnecessary because it is unlikely that barchan dunes actually get preserved.

Barchans are isolated dunes out away from the big sand seas. They generally occur migrating across flat equilibrial surfaces of rock, gravel and sand. It is easy to drive around a barchan, take a few pictures and dig some holes with a reasonable certainty of being back in camp in the evening and surviving to write a paper for one of the learned journals.

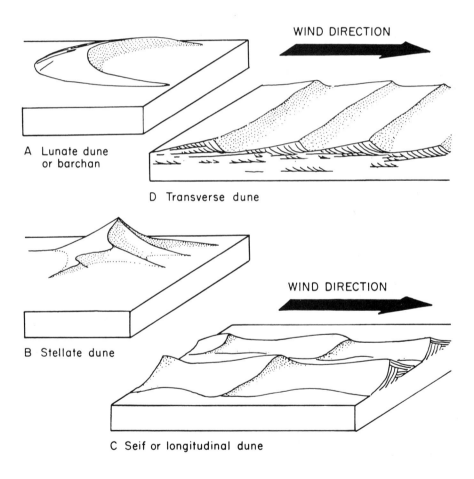

*Figure 3.1* Cartoons of the four common types of dune seen in recent deserts. Probably only the transverse dune is responsible for significant deposition of sand. The other three types are commonly bed forms of sand reworking and transportation across equilibrial surfaces (from Selley, 1982, by courtesy of Academic Press).

The second major dune type is the pyramidal stellate or Matterhorn variety. This is a very beautiful dune to look at, sometimes attaining a height of several hundred feet. Stellate dunes are relatively rare, occurring in tangled patterns within sand seas, or where sand seas but against jebels or escarpments.

The third type of dune is the longitudinal or seif variety. These are generally gregarious, many seifs being aligned parallel to one another and to the prevailing wind direction. Seifs when studied in detail are often seen to be complex bed forms in which the basic longitudinal geometry is punctuated by a series of barchanoid or pyramidal culminations.

Seif dunes are like barchans however, in that they overly flat desert surfaces. They are bedforms in which sand is reworked and transported but from which little net deposition actually takes place.

The transverse dune is the fourth variety to consider. This is seldom described in the literature, but is probably the most important variety. Transverse dunes are not lonely dunes migrating across flat surfaces but gregarious dunes that migrate over the backside of the next dune downwind. It is transverse dunes which actually deposit sands which may be buried and preserved in the stratigraphic record. They are seldom described in the literature because it is difficult to penetrate sandseas and to return to camp for a cool beer in the evening. Essentially transverse dunes are just like the straight-crested megaripples which deposit cross-bedded sands under water.

The interdune area has been defined as 'a geomorphic surface commonly enclosed or at least partially bounded by dunes or other eolian deposits such as sand sheets' (Ahlbrandt and Fryberger, 1981). As already discussed, many interdune areas are equilibrial environments where bed rock is exposed or lag gravels just lie there. In some instances, however, deposition takes place in interdune areas. Ahlbrandt and Fryberger (1981, 1982) recognize three such types: dry, wet, and evaporitic.

Dry interdune deposits consist of massive or flat bedded sand and granules which are generally coarser grained than the adjacent dunes and are bimodally sorted. Wet interdune deposits are rather different. They contain more clay and silt since these particles may be attached to the damp substrate. Similarly, sand may form adhesion ripples as it is blown across the interdune flat. Animal tracks and trails may be preserved, together with Rhizoconcretions (Glennie and Evamy, 1968). Evaporitic interdune deposits develop where sand seas occur adjacent to sabkhas. Nodules, layers and intergranular cements of evaporites are associated with fine-grained flat-bedded terrigenous sediment. Adhesion ripples and shrinkage polygons occur.

Interdune deposits are of significance in two ways. First they may form permeability barriers which may slow down, if not totally inhibit, fluid movement in eolian sands. Secondly, considerable amounts of organic matter have been discovered in recent evaporitic interdune sediments, and it has been suggested that this may act as a petroleum source.

Ancient rocks which have been attributed to eolian deposition are quantitatively insignificant in the sedimentary rocks of the world, but they are widely distributed and range in age from PreCambrian to Recent. Some of the best documented complexes of supposed ancient eolian sediments occur in the western USA. These will now be briefly described and the reasons for attributing them to eolian action stated.

## EOLIAN DEPOSITS OF WESTERN USA: DESCRIPTION

The Colorado Plateau occupies parts of the states of Colorado, Arizona, Utah, and New Mexico in western USA (Fig. 3.2). This region was an important site of intermittent eolian sedimentation through some 150 m.y. from the Pennsylvanian (Upper Carboniferous) to the Upper Jurassic. The regional geology is complicated but can often be resolved due to a combination of good exposure and dissected topography (Fig. 3.3). In general the Cordilleran geosyncline lay to the west and was an area of marine deposition through much of this time. The present Colorado Plateau was then an unstable shelf on which sedimentation took place in a number of basins which were intermittently connected to one another and to the sea. These include the Paradox, San Juan, Kaiparowits, and Black Mesa basins. The stratigraphy of this region is complex

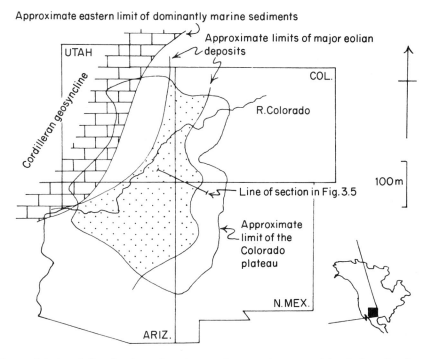

*Figure 3.2* Areal distribution of ancient eolian sandstones of the Colorado plateau, USA.

therefore, with rapid lateral and vertical facies changes due to both tectonic and sedimentary causes. Three main facies can be recognized in the Pennsylvanian to Jurassic rocks of the Colorado Plateau:

(1) Easterly thinning sheets of limestones, shales, dolomites, and evaporites (thought to be marine).

(2) Wedges and fans of red sandstones, siltstones, and conglomerates, generally best developed in the east, e.g. the Moenkopi and Morrison Formations (thought to be fluviatile).

(3) Red and white sandstones with irregular sheet geometries, e.g. the Entrada, Navajo, Wingate, Weber, De Chelly, and Coconino Formations (thought to be eolian).

The supposed eolian sandstones are interbedded with the other two facies with interfingering but generally abrupt margins. They are irregular in plan and seldom greater than 60 m thick though some examples like the Navajo Sandstone approach 300 m Lithologically these deposits are exclusively sand-grade sediment. Petrographically they are protoquartzites with minor amounts of feldspar, mica, chert, and red ferruginous clay. Cementation is by quartz and

*Figure 3.3* Isolated butte illustrating good exposure of the Colorado plateau. Monument Valley, Arizona. Massive Wingate eolian sandstone (Trias-Jurassic) forms a steep cliff above a talus slope with ledges of fluviatile Chinle sandstones and shales (Trias). Photo by courtesy of W. F. Tanner.

calcite. In one or two exceptional cases, as in the upper part of the Todilto Formation, wind-blown gypsum sands occur. The grade of the sandstones range from very fine to coarse, but it is mainly fine. Sorting is moderate to good, occasionally bimodal, with coarse grains in a fine sand. Rounding of grains is moderate to good.

Cross-bedding is the characteristic sedimentary structure of these rocks (Fig. 3.4). Both tabular planar and trough sets are present. Set heights vary from 1–60 m and troughs range in width from 1–60 m. Individual foresets dip at between 20–30° and are generally curved at the base. Low angle backsets also occur sub-parallel to erosion surfaces. Penecontemporaneous deformation of bedding is sometimes present. Some of the fine sandstones show long wavelength low amplitude asymmetrical ripples with R.I.s between 20 and 50 (the ripple index (R.I.) is calculated by dividing the wavelength by the amplitude).

*Figure 3.4* Eolian cross-bedding in Entrada Formation (Jurassic), Church rock. New Mexico. Photo by courtesy of W. F. Tanner.

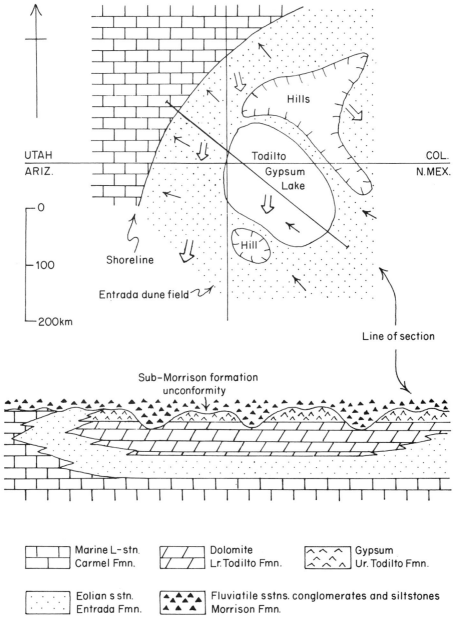

UTAH
ARIZ.

COL.
N.MEX.

UTAH / ARIZ. ... COL. / N.MEX.

Hills

Todilto
Gypsum
Lake

Hill

Shoreline

Entrada dune field

0

100

200km

Line of section

Sub-Morrison formation
unconformity

| | Marine L-stn. |
| | Carmel Fmn. |

| | Eolian s stn. |
| | Entrada Fmn. |

| | Dolomite |
| | Lr. Todilto Fmn. |

| | Fluviatile sstns. conglomerates and siltstones |
| | Morrison Fmn. |

| | Gypsum |
| | Ur. Todilto Fmn. |

*Figure 3.5* Reconstruction of Entrada sandstone palaeogeography. Cross-section about 100 m thick. Modified from Tanner, 1965, Fig. 7. Reproduced from the *Journal of Sedimentary Petrology*, by courtesy of the Society of Economic Paleontologists and Mineralogists. Thin arrows: slope controlled seaward flowing fluvial palaeocurrents. Thick arrows: slope independent eolian palaeocurrents.

Dip directions of cross-bedding show wide scatters within one outcrop. Throughout the majority of Pennsylvanian to Upper Jurassic time they indicate transport generally to the south, with some variation to southeasterly and southwesterly directions.

These sandstones are devoid of fossils except for occasional footprints attributed to terrestrial quadrupeds and bipeds, largely dinosaurs.

## EOLIAN SANDSTONES OF WESTERN USA: DISCUSSION

There are two main lines of argument for the eolian origin of these sands. Some arguments are negative, suggesting that these are not water-laid sands, others are positive, suggesting that they are wind-blown.

It is unlikely that these are water-laid sands because of the absence of pebbles, which are generally too heavy to be wind-blown, and of clay, which is generally too light to come to rest on a windy earth. There is no sign of an aqueous biota either marine or non-marine. Conglomerate-lined channels suggestive of running water are absent.

Positively in favour of an eolian origin for these sandstones is their close comparison with Recent desert dunes. Points in common are the dominance of sand-grade sediment which is generally well sorted, matrix free, and well-rounded. Cross-bedding of such vast height is unknown in Recent aqueous sediments, but is known from Recent terrestrial dunes. Ripple indices >15 are recorded from Recent eolian ripples whereas aqueous ripples have R.I.s <15 (Tanner, 1967). Recent dune country is generally lifeless, such plants and animals as live and die there are desiccated and destroyed by the shifting sands. The persistent southerly direction of sand transport is perpendicular to the (general) westerly palaeoslope of the Colorado Plateau and locally can be seen trending oblique to slope-controlled palaeocurrents of the associated fluviatile red beds (Fig. 3.5).

Some doubt has been expressed as to the eolian origin of these sands. It has been pointed out that the large scale cross-bedding, the well-rounded, well-sorted texture and frosted grain surfaces of these sediments could also have formed in a marine shelf sea. Log-probability sorting curves of the sands are comparable to those of modern tidal-current environments. Green pellets have also been recorded, and interpreted as glauconite of marine origin (Stanley *et al.*, 1971; Visher, 1971; Freeman and Visher, 1975).

It is difficult to pontificate in such a controversy, especially without field experience of the rocks in question. Two points can be made, however. First, this dilemma points up the absence of any clear criteria for recognizing eolian deposits. Second, on a broad shelf where a tidal shallow sea transgresses and regresses a desert, eolian sands may inherit aqueous textural characteristics and vice versa.

Due to the wealth of stratigraphic and sedimentologic data available it is possible to reconstruct the palaeogeography of the ancient Colorado Plateau area in some detail. In Pennsylvanian and early Permian time dunes extended intermittently along a coastal plain separating the Rocky Mountain geosynclinal seas from alluvial piedmont fans to the east. There are no known eolian sands of Upper Permian to Middle Triassic age.

In Late Triassic and Early Jurassic time dune fields developed in inland basins away from the sea. Middle Jurassic eolian sandstones are unknown. In Upper Jurassic time both coastal and interior basin dune fields developed (Fig. 3.6). Tanner (1965) has produced a particularly elegant detailed study of the Upper Jurassic palaeogeography of the 'Four Corners' region of the Colorado Plateau. Marine deposits with southwesterly directed longshore palaeocurrents interfinger eastwards with the Entrada sandstone. This was deposited by a complex of southeasterly migrating coastal dunes which, for a time, separated the open sea from the hypersaline deposits of the Todilto Formation lake. Tongues of fluvial sediment within the eolian Entrada sandstone show northwesterly directed cross-bedding indicating the direction of the palaeoslope as the river flowed down to the sea (Fig. 3.5).

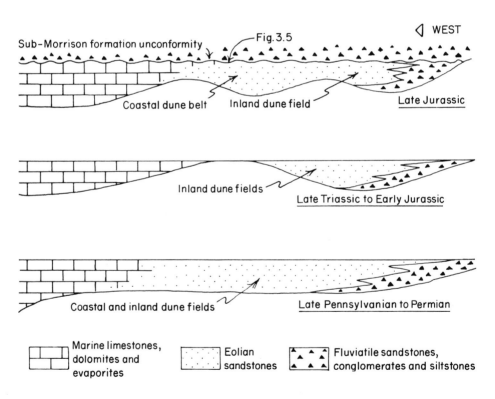

*Figure 3.6* Diagrammatic cross-sections illustrating Colorado Plateau eolian sandstone palaeogeographies.

# GENERAL DISCUSSION OF EOLIAN SEDIMENTS

Because the processes of eolian and aqueous sedimentation are so similar these deposits are extremely difficult to differentiate. Even the classic Mesozoic eolian sands of the Colorado have been subjected to attempts to reinterprete them as we have just seen. There is no single criterion which unequivocally indicates an eolian origin to a sandstone. Only an assemblage of criteria may suffice, backed by the absence of criteria which positively demonstrate a subaqueous environment. Subsurface diagnosis will now be reviewed under the five headings: geometry, lithology, palaeontology, sedimentary structures and palaeocurrent pattern.

There is no distinctive shape to eolian formations. The original dune morphology of a sand sea is generally planed out before sediments of another facies are deposited. Basically a blanket is the final geometry. The environments of adjacent facies may be significant. Because deserts are equilibrial environments sand dunes often migrate across unconformities, cemented by ferrocrete, calcrete or silcrete and overlain by a veneer of pebbles which may include dreikanter, fierkanter and occasionally even the legendary funfkanter. Eolian sands may be expected to pass laterally into braided alluvial outwash sands, playa lake shales and evaporites or fanglomerates. There is no significant seismic response. Because of their uniform sandstone composition there are no significant internal reflectors and, as already discussed, dune palaeotopography is unlikely to be preserved.

According to geological folklore eolian sands are well-rounded, well-sorted orthoquartzites with grain surfaces which are frosted when examined by a microscope, and show diagnostic markings under a scanning electron microscope. Statistical analysis of grain-size curves can differentiate recent eolian sands from subaqueous ones.

Alas, it is not quite so simple. As already discussed granulometric analysis, and surface texture are dangerous tools to use in environmental diagnosis because of the problems of inheritance and reworking, together with the difficulty of disaggregating lithified sandstone, and of surface textures being modified by solution and cementation. It is true that most Saharan and Arabian desert dunes are well-rounded, well-sorted orthoquartzites. But so they would be regardless of their present environment. In both deserts their recent dune sands are poly-poly-polycyclic. They have been through many cycles of weathering, erosion, deposition, transportation and lithification. One will expect these sands to be texturally and mineralogically mature, regardless of their present environment.

In point of fact though most eolian sands are texturally mature, they are often mineralogically very immature.

Well rounded feldspar grains occur in the recent Sahara and in the North Sea Permian dune sands. There are black dunes of volcaniclastic sand in Iceland and around Tertiary lava flows of the Sahara and elsewhere.

Many geologists have proposed textural criteria to distinguish wind-blown from water-laid sand. It is widely accepted that eolian sands are better sorted than aqueous ones. This is generally true, but there is no arbitrary dividing limit. The classic eolian Triassic Bunter Sandstones of England have a Trask sorting coefficient of 1·41 (Shotton, 1937). This is not so well-sorted as a beach sand cited by Krumbein and Pettijohn (1938, p. 232) which has a value of 1·22. It is widely accepted that eolian sands are positively skewed, with a tail of fines (Folk, 1966, p. 88, Folk, 1971). This factor can generally be used to distinguish them from beach sands, but not from fluvial ones (Friedman, 1961, p. 514). Though, as discussed on p. 10, textural criteria have been developed to distinguish eolian sands from aqueous ones in Recent sediments it is hard to use these in ancient deposits for technical reasons (see also Bigarella, 1973). Eolian sands are widely believed to be very well-rounded and certainly experiments show that wind is much more efficient at rounding quartz sand than is running water (Kuenen, 1960). It is important to note, however, the polycyclic history of the eolian sandstones of the Colorado Plateau, the Permo-Trias of Europe, and the present Sahara. One would expect these sands to be well-rounded and well-sorted, whether they were eolian or not.

Eolian sand grains show a frosted, pitted surface under the optical microscope and, under the higher powers of the electron microscope, show a variety of characters which can be used to distinguish them from sands subjected to aqueous and glacial action (Krinsley and Funnel, 1965). All these textural features of eolian sands must be carefully interpreted; remembering that dunes may be re-worked by running water, and alluvial fans may dry out and be re-worked by wind. Water-laid deposits may therefore inherit eolian textural parameters and vice versa. For this same reason wind-faceted pebbles (ventifacts, dreikanters) should be interpreted with care.

Many recent beaches and barrier islands are capped by eolian dunes. These sands often contain marine shell debris and some are indeed composed of nothing but carbonate grains. Coasts around the world are rimmed by Pleistocene limestones with spectacularly large cross-bedding. These are variously termed Bahamite (in the Bahamas) Eolianite or Miliolite (in the large Gulf between Iran and Arabia). Geologists have had many arguments about the eolian or marine origin of these rocks. Limestones may of course become dolomitised, so we must expect eolian dolomites to exist in the subsurface. The problem is worse than that. Some recent eolian dunes are composed of gypsum sands. These occur where sabkhas are deflated and gypsum and crystals eroded and abraded. Recent examples are known from both coastal dunes, like those of Abu Dhabi, and inland dunes, like those in White Sands National Park, New Mexico. The analogous Jurassic Todilto dunes have just been mentioned.

When buried and heated of course, gypsum dehydrates to anhydrite. We must not be surprised therefore if we encounter cross-bedded anhydrite in the subsurface.

Not all eolian deposits are of sand grade. The Pleistocene loess silt deposits of North American, Europe, and China are widely interpreted as due to wind action in arid zones around the ice caps (Smalley, 1975).

It is apparent therefore that it may be difficult to use lithology as a diagnostic criterion of eolian deposition. In a simple world we might look for well rounded, well-sorted medium to fine-grained sand of diverse mineralogical composition, ranging from quartzite and arkose to limestone, dolomite or evaporite. One would not expect to find pebbles or shale clasts in these sands. A red ferric staining should be anticipated, indicating early diagenesis above the water table as discussed on p. 71.

From the preceeding account of eolian sand lithology it is apparent that these deposits may contain, and indeed be totally composed of, a diverse marine fauna. One would however expect the fossils to be extremely fragmented, abraded and rounded.

The presence of abundant bioturbation would be one criteria which would serve to differentiate marine shoal sands from eolian ones; the problem being that these two facies are otherwise remarkably similar in lithology, texture and sedimentary structures. Marine vertical burrows should however be differentiated from rhizoconcretions. These are pipes of carbonate or gypsum cemented sand which form around plant roots (Fig. 3.7). As the plants draw moisture from

*Figure 3.7* Photograph of rhizoconcretions in continental red Triassic sandstone, Sidmouth, England.

the sand salts are precipitated around the rhizomes (Glennie and Evamy, 1968). Rhizoconcretions are not of course unique to eolian sands but also occur in fluvial sands in arid environments. Intraformational conglomerates of fragmented concretionary tubes also occur.

Sedimentary structures have often been used to distinguish wind-blown from water-laid sediment. As already noted (p. 89) long, low ripples have been recorded from wind-blown sediments and Tanner (1967) states that an R.I. of >15 can be used to distinguish wind ripples from those due to water, except for swash zone ripples on beaches. This writer has also empirically derived a number of other statistical criteria for distinguishing eolian and aqueously formed ripples. One might have supposed that contorted sand bedding was restricted to aqueous deposits. Unfortunately this is not so. McKee (1966, p. 69) has recorded it in Recent dunes. It occurs in the Colorado Plateau eolian sandstones and is particularly widespread in the Navajo Formation (Stokes, 1961, p. 158).

Very thick cross-beds are often held to be restricted to eolian sands (Fig. 3.8). However, insufficient is known of Recent marine sand bars to give the maximum set height of water-formed cross-bedding. Therefore no arbitrary

*Figure 3.8*  Photograph of large curvaceous cross-bedding in red Permian sandstone, Dawlish, England. Many geologists accept this style of cross-bedding as characteristic of eolian deposition.

limit between eolian and aqueous set heights may be given. Eolian foresets are often no higher than water-laid sets, and only a fraction of the height of the dune in which they occur (see McKee, 1966, Fig. 8).

Eolian dune foresets are often said to be asymptotically curved towards their bases, and to have steeper inclinations than water-laid foresets. An angle of about 30° is often quoted as critical for distinguishing water-laid from wind-blown foresets. Angles in excess of this figure have been widely recorded from Recent dunes and supposed ancient ones (e.g. McKee, 1966; Mackenzie, 1964; McBride and Hayes, 1963; Laming, 1966; Poole, 1964). On the other hand Shotton (1935) recorded a mean value of only 24° for the Triassic Bunter dune sands of England and Tanner (1965) recorded dips between 21 and 26° as common in the Entrada sandstone of the Colorado Plateau. Care should also ·be taken in studying foreset inclination data from ancient rocks since it is hard to determine the horizontal datum of the time of their formation. Major bedding surfaces generally represent old erosional slopes of the dune surface, and could seldom have been horizontal (see also Potter and Pettijohn, 1963, pp. 78, 80, and 86).

A further problem of eolian sandstones is the determination of their transport directions. Recent dunes show extremely varied morphologies. Transverse dunes whose foresets consistently dip downwind are generally quantitatively subordinate to barchan (lunate), seif (longitudinal), and stellate dunes. These types have much more complex geometries and correspondingly more varied foreset orientations. Barchan foresets curve through an arc of about 180°, while seif dunes show bipolar forest dip orientations perpendicular to mean wind direction (McKee and Tibbits, 1964) (Fig. 3.9). Despite these complexities ancient eolian deposits have produced consistent palaeocurrent trends, though with predictably high scatters of readings. Thompson (1968), in an extremely elegant account of the three-dimensional geometry of Triassic sand dunes in the English Midlands, measured many unimodal dip patterns. Similar results have been obtained from dipmeter logs of Permian Rotliegende dunes of the North Sea (Glennie, 1972). Regional studies have been used in attempts to reconstruct palaeoclimates, particularly the past distribution of high pressure air cells (Shotton, 1956; Opdyke, 1961).

A further problem of eolian palaeocurrents in that they are not slope-controlled like many (but not all) aqueous currents. As we have seen (p. 90) this has been demonstrated in the Entrada sandstone of the Colorado Plateau (Tanner, 1965). In an elegant study of Permotrias desert deposits of southwest England, Laming has shown (1966) how sediment was alternately flushed down alluvial fans by floods and blown upslope again by wind.

In conclusion this discussion shows that it is not easy to recognize ancient wind-blown sediments. This is basically because wind and water both transport and deposit sand in ripples and dunes. Many textural and structural criteria have been proposed to distinguish the results of these processes; few, if any, are foolproof. Recognition of an ancient eolian sediment must be based on a

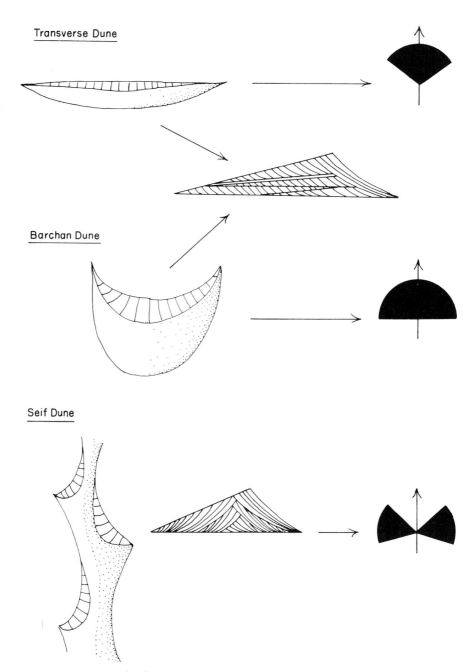

Transverse Dune

Barchan Dune

Seif Dune

*Figure 3.9* Diagrams illustrative of the relationship between dune morphology and cross-bedding orientation shown plotted on azimuths. Wind blows up the page. Barchan and seif dunes are generally ephemeral bed forms which are seldom preserved. Net eolian sedimentation actually appears to come from transverse dunes.

critical evaluation of all the available data. It may help to distinguish criteria which weigh against an aquatic origin separately from those positively in favour of eolian deposition. Finally it often helps to ask the question: 'If this facies is not eolian, what else could it be?'

## ECONOMIC SIGNIFICANCE OF EOLIAN DEPOSITS

Eolian sandstones are potentially of high porosity and permeability because they are typically well-rounded, well-sorted, and generally only lightly cemented. Regional permeability is likely to be good due to absence of shale interbeds. Because of these features eolian sandstones can be important aquifers and hydrocarbon reservoirs. In general, however, eolian sandstones are rated as poor hydrocarbon reservoir prospects. This is because they frequently lie within continental basins far from marine shales which could be potential source rocks. Exceptions to this general rule are found where eolian sandstones are brought into contact with organic-rich source beds by unconformities or by tectonic movement. Some of the North Sea gas fields occur within Permo-Triassic dune sands which unconformably overlie Upper Carboniferous (Pennsylvanian) Coal Measures (Marie, 1975). Deep burial of coal beneath the younger sediments may have caused gas to be expelled during devolatization (Eames, 1975).

In Utah and Wyoming several oil fields produce from some of the Triassic and Jurassic eolian sands described earlier in this chapter. Here, within the foothills of the Rocky Mountains, several of the eolian formations have been thrust over Cretaceous organic-rich shales (Lamb, 1980; Lindquist, 1983).

The distinction of eolian from water-laid sediment can be important where the latter are mineralized and the former are barren. This situation is found in the distribution of Uranium mineralization of the Colorado Plateau (p. 69) and in the Zambian Copper belt (p. 163).

## SUB-SURFACE DIAGNOSIS OF EOLIAN DEPOSITS

From the preceding account it will be apparent that the unequivocal diagnosis of eolian deposits from boreholes is not easy.

Geometry of eolian formations is irregular and lithology is varied. Though the classic eolian sandstone may be a pink well-sorted, well-rounded, medium-fine grained quartzite, other eolian sediments include evaporites, bioclastic limestones and red loessic siltstones.

Perhaps the most diagnostic feature is a clean low A.P.I. gamma log with occasional upward asymmetric sawtooth patterns. The corresponding dip motif is a blue pattern in which the upward increase in dip occurs opposite to the short shaley interval of the gamma curve. This corresponds to the sub-horizontal toeset of the dune on which clay and mica settle out. This passes up into the clean gamma interval with uniform high angle foreset dips (Fig. 3.10). Such eolian sediments tend to be palaeontologically barren.

The Permian Rotliegende of the southern North Sea is probably the best documented example of sub-surface facies analysis of an inland arid basin in

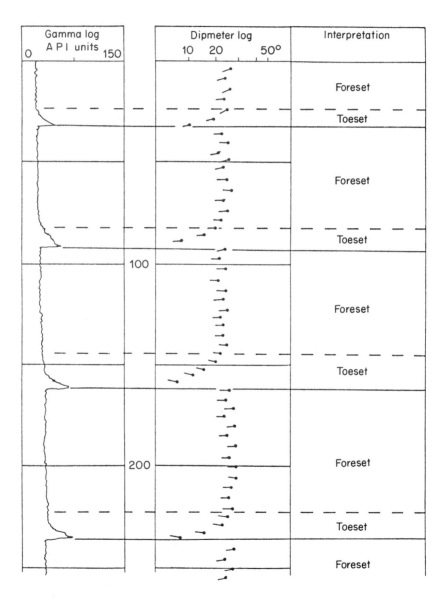

*Figure 3.10* Gamma and dipmeter log motifs for a borehole through eolian dune deposits.

which wireline log interpretation, aided by cores, enables the differentiation of fluvial, eolian and sabkha playa-lake deposits (Glennie, 1972; Marie, 1975; and Van Veen, 1975).

# REFERENCES

The account of the Colorado Plateau eolian sandstones was based on:

Baars, D. L., 1961. Permian Blanket Sandstones of the Colorado Plateau. In: Geometry of Sandstone Bodies. (Eds J. A. Peterson and J. C. Osmond) *Amer. Assoc. Petrol. Geol. Symposium*, pp. 179–207.
Freeman, W. E., and Visher, G. S., 1975. Stratigraphic analysis of the Navajo Sandstone. *J. Sediment. Petrol.*, **45**, 651–8.
Haun, J. D., and Kent, H. C., 1965. Geological history of Rocky Mountain Region. *Bull. Amer. Assoc. Petrol. Geol.*, **49**, 1781–800.
Lessentine, R. H., 1965. Kaiparowits and Black Mesa Basins: Stratigraphic synthesis. *Bull. Amer. Assoc. Petrol. Geol.*, **49**, 1997–2019.
Opdyke, N. D., 1961. The paleoclimatological significance of desert sandstone. In: *Descriptive Paleoclimatology*. (Ed. A. E. M. Nairn) Interscience, N. Y. pp. 45–59.
Poole, F. G., 1964. Paleowinds in the Western United States. In: *Problems of Paleoclimatology*. (Ed. A. E. M. Nairn) Interscience, N.Y. pp. 390–405.
Stanley, K. O., Jordan, W. M., and Dott, R. H., 1971. New hypothesis of Early Jurassic Paleogeography and sediment dispersal for western United States. *Bull. Amer. Assoc. Petrol. Geol.*, **55**, 10–19
Stokes, W. L., 1961. Fluvial and eolian sandstone bodies in Colorado Plateau: In: Geometry of Sandstone Bodies. (Eds J. A. Peterson and J. C. Osmond) *Amer. Assoc. Petrol. Geol. Symposium*, pp. 151–78.
Tanner, W. F., 1965. Upper Jurassic Paleogeography of the Four Corners Region. *J. Sediment. Petrol.*, **35**, 564–74.
Visher, G. S., 1971. Depositional processes and the Navajo sandstones. *Bull. Geol. Soc. Amer.*, **82**, 1421–4.


Other references cited in this chapter are:


Ahlbrandt, T. S., and Fryberger, S. G., 1981. Sedimentary features and significance of interdune deposits. In: *Recent and Ancient non-marine depositional environments: models for exploration*. (Eds F. G. Ethridge and Flores, R. M.) Soc. Econ. Pal. & Min. Sp. Pub., **31**, 293–314.
—— and ——, 1982. Eolian deposits. In: *Sandstone Depositional Environments*. (Eds P. A. Scholle and D. Spearing) Amer. Assoc. Petrol. Geol. Mem., **31**, 49–86.
Bagnold, R. A., 1941. *Physics of Blown Sand and Desert Dunes*. London, Methuen, p. 265.
Bigarella, J. J., 1972. Eolian environments, their characteristics, recognition and importance: In: Recognition of Ancient sedimentary environments. (Ed. J. K. Rigby and W,. K. Hamblin) *Soc. Econ. Pal. Min.* Sp. Pub., No. 16, pp. 12–62.

Bigarella, J. J., 1973. Textural characteristics of the Botucatu sandstone. *Boletin Paranaense de Geociencies*, **31**, 85–94.

Brookfield, M. E. and Ahlbrandt, T. S., 1983. *Eolian Sediments and Processes.* Elsevier, Amsterdam, p. 660.

Eames, T. D., 1975. Coal Rank and Gas Source Relationships — Rotliegendes Reservoirs. In: *Petroleum and the Continental Shelf of northwest Europe* (Ed. A. W. Woodland) Applied Science Pub., London. p. 191–210.

Folk, R. L., 1966. A review of grainsize parameters. *Sedimentology*, **6**, 73–93.

——, 1971. Longitudinal dunes of the northwestern edge of the Simpson Desert, Northern Territory Australia, 1, Geomorphology and grain size relationships. *Sedimentology*, **16**, 5–54.

Friedman, G. M., 1961. Distinction between dune, beach and river sands from their textural characteristics. *J. Sediment. Petrol.*, **31**, 514–29.

Glennie, K. W., 1970. *Desert sedimentary environments.* Elsevier, Amsterdam, p. 222.

——, 1972. Permian Rotliegendes of northwest Europe interpreted in the light of modern sediment studies. *Bull. Amer. Assoc. Pet. Geol.*, **56**, 1048–71.

Glennie, K. W. and Evamy, B. D., 1968. Dikaka: Plant and plant root structures associated with eolian sand. *Paeogeog. Palaeoclimatol. and Palaeoecol.*, **4**, 77–87.

Krinsley, D. H., and Funnel, B. M., 1965. Environmental history of sand grains from the Lower and Middle Pleistocene of Norfolk, England. *Quart. J. Geol. Soc. London.* **121**, 435–61.

Krumbein, W. C., and Pettijohn, F. J., 1938. *Manual of Sedimentary Petrography.* Appleton-Century-Crofts, Inc., N.Y., p. 549.

Kuenen, Ph., 1960. Experimental abrasion: 4, Eolian action. *J. Geol.*, **68**, 427–49.

Lamb, C. F., 1980. Painter Reservoir Field: Giant in the Wyoming Thrust Belt. In: *Giant Oil and Gas Fields of the Decade 1968–78.* (Ed. M. T. Halbouty) Amer. Assoc. Petrol. Geol. Mem. 30, 281–88.

Laming, D. J. C., 1966. Imbrication, Paleocurrents and other sedimentary features in the Lower New Red Sandstone, Devonshire, England. *J. Sediment. Petrol.*, **36**, 940–57.

Lundquist, S. J. 1983. Nugget Formation Reservoir Characteristics Affecting Production in the Overthrust Belt of Southwestern Wyoming. *Jl. Pet. Tech.*, **35**, 1355–65.

Mackenzie, F. T., 1964. Bermuda Pleistocene eolianites and paleowinds. *Sedimentology*, **3**, 52–64.

McBride, E. F., and Hayes, M. O., 1962. Dune cross-bedding on Mustang Island, Texas. *Bull. Amer. Assoc. Petrol. Geol.*, **46**, 546–52.

McKee, E. D., 1966. Structures of dunes at White Sands National Monument, New Mexico (and a comparison with structures of dunes from other selected areas). *Sedimentology*, **7**, No. 1, Sp. Issue, p. 69.

McKee, E. D. 1979. A Study of Global Sand Seas. *U.S. Geol. Surv. Prof. Pap.* 1052, p. 423.

McKee, E. D., and Tibbitts, G. C., 1964. Primary structures of a seif dune and associated deposits in Libya. *J. Sediment. Petrol.*, **34**, 5–17.

Marie, J. P. P., 1975. Rotliegendes stratigraphy and Diagenesis. In: *Petroleum and the continental shelf of northwest European*, (Ed. A. W. Woodland) Applied Science Pub., London. p. 202–12.

Potter, P. E., and Pettijohn, F. J., 1963. *Paleocurrents and Basin Analysis.* Springer-Verlag, Berlin, p. 296.

Sanders, J. E., and Friedman, G. M., 1967. Origin and occurrence of Limestones. In: *Carbonate Rocks*, Part A (Ed. G. V. Chilingar, H. J. Bissell, and R. W. Fairbridge) Elsevier, Amsterdam. pp. 169–265.

Shotton, F. W., 1937. The Lower Bunter Sandstones of North Worcestershire and East Shropshire. *Geol. Mag.*, **74**, 534–53.

Smalley, I. J., 1975. *Loess: lithology and genesis.* J. Wiley & Sons, p. 429.

Tanner, W. F., 1967. Ripple mark indices and their uses. *Sedimentology*, **9**, 89–104.

Thompson, D. B., 1969. Dome-shaped Aeolian dunes in the Frodsham Member of the so-called 'Keuper' sandstone formation (Scythian-Anisian: Triassic) at Frosham, Cheshire, (England). *Sedimentary Geology*, **3**, 263–89.

Van Veen, F. R., 1975. Geology of the Leman gas field. In: *Petroleum and the continental shelf of northwest Europe*, Vol. 1. (Ed. A. W. Woodland) Applied Science Pub., London. pp. 223–32.

Wilson, I. G., 1972. Aeolian bedforms — their development and origins. *Sedimentology*, **19**, 173–210.

Wilson, I. G. 1973. Ergs. *Sed. Geol.*, **10**, 76–106.

# 4    Lake deposits

## INTRODUCTION

Lakes are landlocked bodies of non-marine water. Recent examples may be hundreds of kilometres long, as are the Canadian Great Lakes, and of widely varying depths. They may be permanent bodies of water or only of short duration. For example, the playa lakes of central Australia last for only weeks or months every few years after torrential rainstorms. Lakes occur in mountainous regions, as in the Alps, and on low-lying continental platforms, like Lake Chad. Their waters range from fresh in temperate and periglacial climates to hypersaline in arid regions. Lakes may be caused by tectonic subsidence and faulting, by glacial erosion, and by damming due to ice, lava, and, recently, man.

Because of their widely variable environmental settings Recent lacustrine sediments are of many kinds. We must expect ancient lacustrine deposits to show a similar diversity. It is beyond the scope of this book to examine all the different kinds and, after briefly reviewing recent lakes, one case history must suffice. This is followed however by a review of some other ancient lake deposits. Both recent and ancient lakes have been intensively studied for economic as well as aesthetic reasons. Reviews of recent lakes and their relation to ancient ones have been given by Collinson (1978), Matter and Tucker (1978), Picard and High (1981), Fouch and Dean (1982), and Dean and Fouch (1983).

## RECENT LAKES

The chemistry, physics and biology of modern lakes have been extensively studied. But rather less is known about their sedimentology. Most of this work has concentrated on the permanent lakes of temperate climates (Hutchinson, 1957; Reeves, 1967 and Lerman, 1978). The sediments of recent ephemeral lakes have been studied by Reeves, 1972, and Neal, 1975).

These studies have shown that the two main controlling parameters of lacustrine sedimentation are the degree of aridity and the amount of sediment input. The latter is controlled partly by climate (temperature, and the type, frequency and amount of precipitation) and partly by the relief and rock type of the hinterland. These variables have been used as a basis for classifying lakes by Kukal, 1971; Visher, 1974; Picard and High, 1972. Figure 4.1 presents a simple classification of lakes that can be recognized both on the surface of the earth today and in the stratigraphic record. Around the lake shores rivers may build out deltas from which wind generated currents can transport sand in beaches and bars. Where there is no influx of terrigenous detritus swamps may deposit peat. On some lake shores lime is deposited. This is largely of biogenic origin, ranging from algal reefs to detrital sands of mollusc shells and algal debris

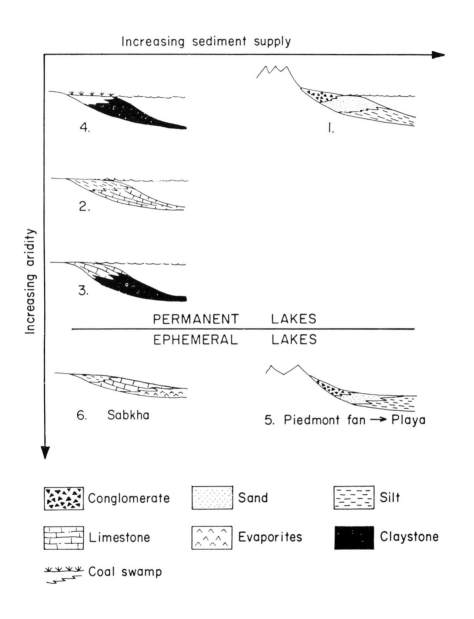

Increasing sediment supply

Increasing aridity

4.

2.

3.

1.

PERMANENT LAKES

EPHEMERAL LAKES

6. Sabkha

5. Piedmont fan → Playa

Conglomerate    Sand    Silt

Limestone    Evaporites    Claystone

Coal swamp

*Figure 4.1* Cartoons to illustrate the various types of lake classified according to rate of sediment supply and the degree of aridity.

and even to oolite sands. Turbidity currents may transport detritus from the shoreline far out into the basin centre where the finest sediment comes to rest. This may include not only terrigenous detritus but also transported lime and humic organic matter. Extensive algal growth within the shallow photic zone in the centres of lakes may give rise to the deposition of sapropelic muds.

In many lakes a stratification of water develops, with warm shallow water overlying cooler denser water. The activity of phytoplankton in the upper layer generates oxygen which supports many other forms of life. If a lake has a density stratification then the oxygen in the lower layer becomes depleted, and cannot be renewed because it is too dark for photosynthesis. In this situation the lower layer becomes anoxic and organic matter settling from above may be preserved. These two layers are referred to as the Epilimnion and the Hypolimnion respectively (Fig. 4.2).

Intermittent storms may mix the two water masses. This often leads to catastrophic mortality of organisms in the upper layer whose bodies may then

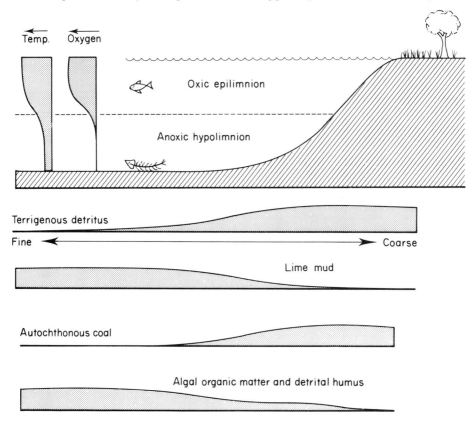

*Figure 4.2* Cartoon to show the relationship between sediment type and some relevant physical and chemical parameters in a lake with a thermally induced density stratification.

be preserved on the lake floor as anoxic conditions reestablish themselves in the Hypolimnion. This is why many ancient lake deposits contain oil shales and petroleum source beds in their centres. Regular repetition of this process imparts a cyclic lamination to the lake sediments. Cyclicity is also commonly seen in glacial lakes where laminae of terrigenous detritus from melt-water streams are interlayered with lime mud and organic matter. These regular layers, termed 'varves', are of course annual.

Sedimentation in ephemeral desert lakes has been described by Reeves (1972) and Neal (1975). Lake basins in arid mountainous terrain are often referred to as playas. Red clays and silts in the centre of the lake basin generally pass via alluvial outwash plains into fanglomerates adjacent to mountains. Many playa lakes are only flooded for a few months every twenty years or so as a result of torrential storms and flash floods (it has been remarked that more people are drowned in the Sahara than die of thirst). For the rest of the time the lake basins are dry flat surfaces with intermittent sand dunes around their edges and occasional huge desiccation cracks in their centres.

Some lakes in arid regions are areas of evaporite sedimentation. Recent examples include the Great Salt Lake of Utah, the Dead Sea, and Lake Eyre, Australia.

## THE GREEN RIVER FORMATION (EOCENE) ROCKY MOUNTAINS, USA: DESCRIPTION

At the end of Cretaceous time the Laramide orogeny threw up the Rocky Mountains in a north–south line trending through western central North America. A series of basins developed on the eastern side of these which were infilled by Tertiary continental sediments. During the Eocene Period fluviatile and lacustrine deposits were laid down in parts of Wyoming, Colorado, and Utah. Sandstones, silts, and coals of the Wasatch Formation are overlain by the lithologically diverse Green River Formation. This is one of the largest and best documented examples of ancient lacustrine deposits in the world. Sedimentation of this formation took place essentially in two areas (Fig. 4.3). The largest of these, Lake Uinta, covered some 23 000 sq. km of northeastern Utah with a depocentre in the Uinta basin where over 3000 m of sediment were laid down. This region was separated by a ridge from the smaller lake Gosiute with a depocentre in the Green River basin of southwestern Wyoming.

In geometry therefore the Green River Formation is irregular in plan and thickness, its margins being controlled partly by syndepositional topography and tectonics, and partly by subsequent erosion.

The Green River Formation consists of a wide range of rock types with complex facies relationships and a correspondingly complex stratigraphic nomenclature. Two major facies can be recognized; clastics and carbonates around the margins pass basinward into fine-grained clastics, carbonates, and evaporites. The marginal facies contains rare conglomerates, cross-bedded and

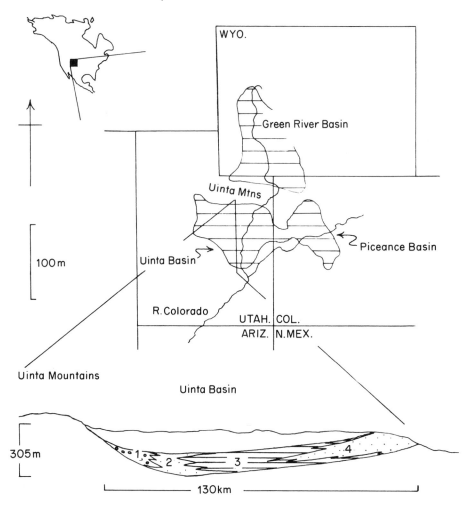

*Figure 4.3* Upper: Present approximate areal extent of Green River Formation. Modified from Haun and Kent, 1965. Fig. 23. From the *Bulletin of the American Association of Petroleum Geologists*, by courtesy of the American Association of Petroleum Geologists. Lower: Cross-section through the Uinta basin showing distribution of Green River facies. 1. Fluvial sand facies. 2. Delta and nearshore sand facies. 3. Oil shale facies. 4. Mud flat deltaic deposits. Based on Osmond, 1965, Fig. 10. From the *Bulletin of the American Association of Petroleum Geologists*, by courtesy of the American Association of Petroleum Geologists.

channelled sandstones, silts, algal stromatolites, dolomites, oolites, and shell beds. This facies interfingers downwards and laterally into the coal-bearing wholly clastic Wasatch Formation. The central parts of the basins consist of thick laminated sequences of marls, fine calcareous sandstones, shales, and oil shales with saline deposits, which include trona (sodium carbonate), towards

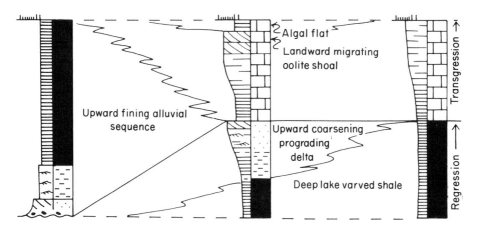

*Figure 4.4* Idealized shoreline cycle in the Green River Formation. The fluvial and shoreline cycles (left-hand and central log) vary in thickness from 2–15 m. The deep water cycles (right-hand log) are generally less than 2 m thick. Based on Picard, and High, 1968, Fig. 4. From the *Journal of Sedimentary Petrology*, by courtesy of the Society of Economic Paleontologists and Mineralogists.

the top. The fine calcareous sandstones show graded bedding; the oil shales show layered couplets less than 1 mm thick which are comparable to the varves of glacial deposits. The lowest part of each couplet consists of very fine calcareous quartz silt, and the upper part is composed of black sapropelic material.

Cyclicity is also developed on a larger scale in the marginal deposits, noticeably on the northeastern flank of the Uinta basin. Each cycle consists of a lower clastic unit which coarsens upwards from shale to sand and thus to an erosion surface towards the basin margin. The erosion surfaces are overlain by upward-fining clastic sequences grading from conglomerate to red shale. This sequence passes upwards and basinwards into oolites and algal carbonates (Fig. 4.4).

The Green River Formation contains an abundant, well-preserved, and diverse biota. This includes freshwater fish, insects, molluscs, ostracods, plankton, pollen, calcareous algae, plant remains, and insects.

In the marginal clastic sediments both cross-bedding and ripple marks have been used to determine palaeocurrents. These indicate sediment transport to and fro up and down local shoreline trends. Palaeocurrent studies have been used to predict the geometries of oil-bearing shoreline sand bodies.

## GREEN RIVER FORMATION: DISCUSSION OF ENVIRONMENT

The preponderance of fine-grained laminated sediment shows that this is an aqueous deposit. The absence of a marine fauna indicates that these waters were non-marine, though terminally hypersaline. By definition, therefore, the Green River Formation originated in a lacustrine environment.

With the wealth of data available, however, it is possible to be much more specific about the origin and history of this lake. The absence of cross-bedding and desiccation cracks in the central facies suggests that deposition took place below wave base in waters that never dried out, at least until the terminal hypersaline phase. The varves have been interpreted as annual events, the lower clastic part indicating an influx of land-derived material during the winter rainy period. The sapropelic layers suggest that organic matter from the surface waters rained down on the lake bed during dry summers when there was low runoff and negligible sediment influx from the land. The preservation of this organic matter, together with perfectly preserved fish skeletons and an absence of benthos, suggest that the deeper parts of the lake were anaerobic and stagnant. Assuming that the varves are annual, knowing their average thickness, and the total thickness of varved sediment, it has been calculated that the lake existed for more than 7 500 000 years. Comparison of the Green River flora with Recent plants indicates that the lake and surrounding hills supported a lush vegetation which grew in a warm humid climate.

In the Uinta basin of Utah the large-scale sedimentary cycles around the margins of the lake have been interpreted as due to transgressions and regressions of the shoreline as the lake waters rose and fell. A drop in level would cause the marginal rivers to downcut to new base levels and carry out the detritus to deposit prograding deltas; hence the upward-coarsening sequences passing basinwards to shale. A rise in lake level would cause the river channels to be choked in their own detritus. Hence the formation of an upward-fining alluvial sequence; the red shale at the top of which suggests the sub-aerial exposure of a flood plain. Simultaneously the drop in clastic supply to the lake would allow nearshore carbonate sedimentation, with oolite banks, shell beds, and calcareous algal mats. These would transgress over the marginal floodplains as the lake rose until the next drop in water level.

For the Green River basin of Wyoming a predominantly playa lake environment has been proposed, in which the oil shale and trona deposits accumulated in shallow lakes surrounded by broad flat playas where the carbonate and terrigenous facies originated.

It can be seen therefore that the Green River Formation is a good example of an ancient lacustrine deposit. Though some of the more romantic interpretations may be open to question there can be no doubt that these rocks record the past existence of a lake of considerable dimensions.

## GENERAL DISCUSSION OF ANCIENT LAKE DEPOSITS

Basically lacustrine sediments can easily be identified by the combination of low energy aqueous deposits and the absence of a marine fauna (Picard and

High, 1972a). However, as seen in the case of the grey facies of the Torridonian, in PreCambrian sediments with no fauna the distinction between lacustrine and marine sediments may be difficult (p. 52). The Green River Formation possesses a variety of characteristics which are found in other ancient lakes. In particular note the lithological diversity. Though the majority of lakes are realms of fine-grained sedimentation these may be either argillaceous, calcareous, or evaporitic. Marginal sediments of lakes may be very coarse and conglomeratic, as in the case of the Newark trough (Triassic) lacustrine sediments which pass laterally into fanglomerates (Portland arkose and Newark conglomerate series).

A second feature typical of ancient lakes shown by the Green River Formation is cyclicity. Varves occur not only in this example but, where they were first defined, in Pleistocene glacial lakes of northern Europe (note also that the flora of the Green River Formation demonstrates that varves are not a criterion of glacial climates). Varves also occur in the lacustrine Lockatong Formation (Triassic) of the Newark trough where they occur within large cycles capped by desiccation cracks indicating regular fluctuations and evaporation of the lake waters (Van Houten, 1964). A detailed account of cycles in lacustrine regimes is given by Duff, Hallam, and Walton (1967). Figure 4.5 shows lamination in Pleistocene lacustrine sediments from the Sahara, and Fig. 4.6 shows some contorted laminated lacustrine sediments from the Dead Sea.

*Figure 4.5*  Rhythmic lamination in Pleistocene lacustrine sediments, Libyan Sahara.

*Figure 4.6* Contorted laminations in Pleistocene Lisan Marls, Dead Sea.

A third feature of some ancient lakes shown by the Green River Formation is the presence of evaporites. These also occur in Recent lakes such as the Dead Sea and the Great Salt Lake of Utah. The source of lacustrine evaporites is often debatable. They may be derived from evaporitic rocks outcropping in the lakes' catchment area. The lake may have formed from a cut-off arm of the sea. Alternatively the salts may have come in the atmosphere from the oceans, being precipitated with rain and concentrated by rivers into ephemeral salt lakes. When attempting to find the source for lacustrine evaporites (where even their lacustrine origin may be in doubt) each case must be judged by its own merits. Thus in the case of the Jurassic Todilto evaporites of the Colorado Plateau it may be significant that they are only a short distance from time equivalent marine strata (p. 90), whereas the Triassic Cheshire salt deposits of northern England, whose origin is in doubt, are far from time equivalent marine limestones and dolomites. Such cases as these need careful observation and interpretation. Even the presence of marine micro-fossils in evaporites is no proof of a marine origin. Holland (1912) records foraminifera being blown 200 miles inland from the Runn of Kutch to be deposited with evaporites in the Great Sambhar Lake of the Rajputana Desert. This is a useful cautionary note on which to end this discussion.

## ECONOMIC ASPECTS

Ancient lakes are of considerable economic importance for the production of petroleum, coal, and evaporite minerals.

There are many ancient lake basins that contain organic-rich shales in their centres which have generated petroleum now trapped in sands around the old lake basin shoreline. The Green River Shale Formation has generated oil in several of the Tertiary basins of Utah and Wyoming (Chatfield, 1972; Randy Ray, 1982). Many other petroliferous lake basins occur in China, Brazil and Australia (Chin Chen, 1980; Guangming and Quanheng, 1982; Gruner and Grunau, 1978 and Evans, 1981).

Many oil shales are of lacustrine origin (Yen and Chilingarian, 1976). The Green River oil shale has been noted already. Other examples occur in the Permo-Carboniferous of New South Wales and in the Carboniferous deposits of 'Lake Cadell' in the Midland Valley of Scotland (Greensmith, 1968).

Important coal fields also occur in ancient fluvio-lacustrine swamps (Falini, 1965). This type of environmental setting is widespread in Permian coal fields of the continents which are interpreted as once having formed Gondwanaland. These include those of the Karro System in Zambia, South Africa, and Zimbabwe; the Damuda System of India, and equivalent strata in Australia and Antarctica. These coal deposits are all of about the same (Permian) age, they often overlie sediments of possible glacial origin and occur in thick cyclic non-marine clastic sequences. The majority of the coals seem to have originated in swamps on alluvial plains peripheral to lakes (Duff, Hallam, and Walton, 1967, pp. 38–43).

The Oligocene Bovey Tracey basin of south Devon is an example of an ancient lake which contains both lignite and kaolin clays (Edwards, 1976).

Other economic minerals formed in lacustrine deposits include iron ore (generally limonite), kiesulguhr (diatomaceous earth) and evaporites.

## SUB-SURFACE DIAGNOSIS OF LACUSTRINE DEPOSITS

From the preceding account of lacustrine deposits it can be seen that they are virtually indistinguishable from marine ones. The main diagnostic criterion is an absence of marine biota (in Phaneorozoic sediments). There are also some differences between the geochemistry of lacustrine and marine evaporites.

## REFERENCES

The description of the Green River Formation was based on the following references:

Bradley, W. H., 1929. Algal reefs and oolites of the Green River Formation. *U.S. Geol. Surv. Prof. Paper* 154-G, pp. 203–23.
——, 1929. The varves and climates of the Green River Epoch. *U.S. Geol. Surv. Prof. Paper* 158-E, pp. 88–95.
——, 1931. Origin and microfossils of the oil shale of the Green River Formation of Colorado and Utah. *U.S. Geol. Surv. Prof. Paper* 168, p. 58.
——, 1931. Non-glacial marine varves. *Am. J. Sci.*, **222**, 318–30.
——, 1948. Limnology and the Eocene lakes of the Rocky Mountains region. *Bull. Geol. Soc. Amer.*, **59**, 635–48.

——, 1973. Oil shale formed in desert environment: Green River Formation, Wyoming. *Bull. Geol. Soc. Amer.*, **84**, 1121–4.

Cole, R. D., and Picard, M. D., 1981. Sulfur-isotope variations in marginal lacustrine rocks of the Green River Formation, Colorado and Utah. In: *Recent and Ancient Non-Marine depositional environments: models for exploration.* (Eds F. G. Ethbridge and R. M. Flores) Soc. Econ. Pal. & Min. Sp. Pub. No. 31. 261–75.

Eugster, H. P., and Surdam, R. C., 1973. Depositional environment of the Green River formation of Wyoming, a preliminary report. *Bull. Geol. Soc. Amer.*, **84**, 1115–20.

Lundell, L. L., and Surdam, R. C., 1975. Playa Lake depositional Green River Formation Piceance Creek Basin Colorado. *Geology*, **3**, 493–7.

Picard, M. D., 1957. Criteria for distinguishing lacustrine and fluvial sediments in Tertiary beds of Uinta Basin, Utah. *J. Sediment. Petrol.*, **27**, 373–7.

——, 1967. Paleocurrents and shoreline orientations in Green River Formation (Eocene), Raven Ridge and Red Wash areas, North-eastern Uinta Basin, Utah. *Bull. Amer. Assoc. Petrol. Geol.*, **51**, 383–92.

——, and High, L. R. Jr, 1968. Sedimentary cycles in the Green River Formation (Eocene), Uinta Basin, Utah. *J. Sediment. Petrol.*, **38**, 378–83.

——, and High, L. R., 1972. Palaeoenvironmental reconstructions in an area of rapid facies changes, Parachute Creek Member of Green River Formation (Eocene), Utah Basin, Utah. *Bull. Geol. Soc. Amer.*, **83**, 2689–708.

Surdam, R. C. and Stanley, K. O., 1979. Lacustrine Sedimentation during the culminating phase of Eocene Lake Gosuite, Wyoming. (Green River Formation). Bull. Geol. Soc. Amer., **90**, 93–110.


Other references cited in this chapter were:


Chatfield, J., 1972. Case History of Red Wash Field, Uinta County, Utah. In: *Stratigraphic Oil and Gas Fields.* (Ed. R. E. King) Amer. Assoc. Petrol. Geol. Mem. 16, 342–53.

Chin Chen, 1980. Non-Marine Setting of Petroleum in the Sungliao Basin of Northeastern China. *J. Pet. Geol.*, **2**, 233–64.

Collinson, J. D., 1978. Lakes. In: *Sedimentary Environments and Facies.* (Ed. H. G. Reading). Blackwell, Oxford, 61–79.

Dean, W. E. and Fouch, T. D., 1983. Lacustrine Environments. In: *Carbonate Depositional Environments.* (Ed. P. A. Scholle, D. G. Bebout and C. H. Moore). Amer. Assoc. Petrol. Geol. Mem. 33, 97–130.

Duff, P. McL. D., Hallam, A., and Walton, E. K., 1967. *Cyclic Sedimentation.* Elsevier, Amsterdam, p. 280.

Edwards, R. A., 1976. Tertiary sediment and structure of the Bovey Basin, south Devon. *Proc. Geol. Assn. London*, **87**, 1–26.

Evans, P. R., 1981. The petroleum potential of Australia. *Jl. Pet. geol.*, **4**, 123–46.

Falini, F., 1965. On the formation of coal deposits of Lacustrine origin. *Bull. Geol. Soc. Amer.*, **76**, 1317–46.

Fouch, T. D. and Dean, W. E. 1982. Lacustrine Environments. In: *Sandstone Depositional Environments.* (Ed. P. A. Scholle and D. Spearing). Amer. Assoc. Petrol. Geol. Tulsa. 87–114.

Greensmith, T., 1968. Palaeogeography and rhythmic deposition in the Scottish oil-shale group. *Proc. U.N. Symp. Dev and Util. Oil Shale* Res. Tallin. p. 16.

Grunau, H. R., and Gruner, U., 1978. Source rock and origin of natural gas in the Far East. *Jl. Pet. geol.*, **1**, 3–56.

Guangming, Z., and Quanheng, Z., 1982. Buried-Hill Oil and Gas Pools in the North China Basin. In: *The Deliberate Search for the Subtle Trap.* (Ed. M. T. Halbouty). Amer. Assoc. Petrol. Geol. Mem. No. 32. 317–36.

Holland, Sir T. H., 1912. The origin of Desert Salt deposits. *Proc. Lpool. Geol. Soc.*, **11**, pp. 227–50.

Hutchinson, G. E., 1957. *A Treatise on Limnology.* J. Wiley, N.Y. 1015 p.

Keighin, C. W. and Fouch, T. D. 1981. Depositional environments and diagenesis of some non-marine Upper Cretaceous Reservoir Rocks, Uinta basin, Utah. In: *Recent and Ancient Non-marine depositional environments: models for exploration.* (Eds F. G. Ethridge and R. M. Flores). Soc. Econ. Pal. & Min. Sp. Pub. 31. 109–25.

Lerman, A. (Ed.), 1978. *Lakes: Chemistry, Geology and Geophysics.* Springer-Verlag, Berlin, p. 363.

Mather, A. and Tucker, M. E. 1978. Modern and Ancient Lake Sediments. *Spec. Pub. Internat. Assn. Sedol.*, **2**, p. 296.

Neal, J. T. (Ed.) 1975. *Playas and dried lakes.* J. Wiley & Sons, New York, p. 411.

Picard, M. D., and High, L. R., 1972a. Criteria for recognizing Lacustrine rocks. In: Recognition of Ancient sedimentary environments. (Ed. J. K. Rigby and W. K. Hamblin) *Soc. Econ. Pal. Min.*, Sp. Pub. No. 16, 108–45.

Picard, M. D. and High, L. R. 1981. Physical Stratigraphy of Ancient Lacustrine Deposits. In: *Recent and Ancient Nonmarine Depositional Environments: Models for Exploration.* (Eds F. G. Ethridge and R. M. Flores) Soc. Econ. Pal. & Min. Sp. Pub. 31. 233–59.

Randy Ray, R., 1982. Seismic stratigraphic interpretation of the Fault Union Formation, western Wind River Basin: example of subtle trap exploration in a nonmarine sequence. In (Ed. M. T. Halbouty). Amer. Assoc. Petrol. Geol. Mem. No. 32. 169–180.

Reeves, C. C., 1967. *Introduction to Palaeolimnology.* Elsevier, Amsterdam, p. 228.

——, 1972. *Playa Lake symposium,* 1970. ICASALS Pub. No. 4. Texas Tech. University, Lubbock, p. 334.

Ryder, R. T., Fouch, T. D., and Elison, J. H., 1976. Early Tertiary sedimentation in the western Uinta Basin, Utah. *Bull. Geol. Soc. Amer.*, **87**, 496–512.

Van Houten, F. B., 1964. Cyclic lacustrine sedimentation. Upper Triassic Lockatong Formation, central New Jersey and adjacent Pennsylvania. *Kansas Geol. Surv. Bull.*, **169**, 497–531.

Visher, G. S., 1965. Use of vertical profile in environmental reconstruction. *Bull. Amer. Assoc. Petrol. Geol.* **42**, 254–71.

Yen, T. F. and Chilingarian, G. V. (Eds), 1976. *Oil Shale.* Elsevier, Amsterdam, p. 292.

# 5    Deltas

## INTRODUCTION

### Shorelines: a general statement

The development of shoreline sedimentary environments and their concomitant facies is a function of many variables, including rate of influx of land-derived sediment, tidal regime, current system, climate, and the relative movements of land and sea. Recent shorelines are sometimes areas of net erosion or net deposition. The former do not concern us since they leave no sediment in the geological record. The product of depositional shorelines, on the other hand, compose a large percentage of the world's sedimentary cover. The nature of shoreline deposits basically reflects the relative strengths of land-derived sediment; and the ability of marine processes to re-distribute it.

*Table 5.1. A classification of shoreline deposits.*

| | |
|---|---|
| I. **Lobate shorelines: deltas**<br>*Mississippi type*, with radiating 'birdfoot' distributary channels.<br>*Nile type*, radiating channels truncated by an arc of barrier islands. | Carbonate<br>facies<br>encroach<br>shorewards |
| II. **Linear shorelines: barrier island and lagoonal complexes**<br>*Terrigenous*, terrigenous sedimentation dominant in the alluvial coastal plain, lagoon, barrier, and open marine environments.<br>*Mixed carbonate: clastic*, terrigenous alluvial plain and lagoon; carbonate barrier and open marine facies.<br>*Carbonate*, limestone open marine and barrier facies. Carbonate facies in lagoonal environment, sometimes dolomitic and evaporitic. Terrigenous coastal plain facies absent or only poorly developed. | <br>Relative<br>increase<br>in land-<br>derived<br>sediment |

The interplay of these two processes produces a continuous spectrum of shoreline types ranging from the complete dominance of the shoreline by terrestrial sediment, to the other extreme where marine processes dominate over a negligible influx of detritus. With declining terrestrial sediment supply, carbonate sedimentary environments migrate closer and closer inshore.

These concepts can be used as a basis on which to arbitrarily classify this continuous spectrum of shorelines into a number of types, as shown in Table 5.1.

This chapter describes lobate (deltaic) shorelines. The next three chapters are concerned with terrigenous, mixed, and carbonate shorelines respectively.

## Recent deltas

Herodotus applied the Greek letter delta (Δ) to the triangular area where the distributaries of the Nile deposit their load into the Mediterranean (*ibid.*, 454 B.C.). This term has gained widespread usage by geographers. More recently Lyell (1854) re-defined a delta as 'alluvial land formed by a river at its mouth'. This second definition is perhaps safer for geologists to use since it lays no stress on the deltaic geometry of such an area. Indeed, many Recent river mouths which are accepted as deltaic are far from triangular in shape (see Shirley and Ragsdale, 1966, pp. 234–51). Nevertheless, it is of extreme economic importance to be able to distinguish lobate deltaic shorelines from linear ones. Fortunately this can often be achieved from the study of a single borehole or surface section without acquiring sufficient data to reconstruct the facies geometry. This is because these two types of shorelines deposit distinctive vertical sequences of grain size and sedimentary structures. Deltas are one of the most studied and written about of all Recent environments. A very brief selected list of this vast literature is given at the end of this chapter.

These studies show that deltas form where a river brings more sediment into the sea than can be re-deployed by marine currents. In such a case the river deposits progressively finer sediment in progressively deeper water as its current velocity diminishes seawards. The channel advances seaward between two raised banks of its own deposits, termed levees. Ultimately the channel lays down so much sediment at its mouth that this is higher than the proximal, landward end of the levees. As the channel chokes in its own detritus, it breaks through the levees on its margin. From the crevasse thus formed a new channel flows seaward between two new levees. Thus an alluvial floodplain builds out to the sea, lobate in plan and wedge-shaped in vertical cross-section. The top of the delta consists of a radiating network of distributary channel sands flanked by levee silts. Finer silts and clays are laid down in the flood basins and lagoons of the interdistributary areas. Swamps often form and are the site of peat-deposition. Seaward of the distributary channel mouth, sediment is deposited on the sub-aqueous delta platform. Rippled interlaminated sand and mud are formed, often with bioturbation. If sedimentation is relatively slow, the platform facies may be re-worked by marine currents. The mud may be winnowed out to leave cross-bedded or flat-bedded sands.

The sub-aqueous delta platform is separated from the prodelta by the delta slope. On this surface most of the mud finally comes to rest. From time to time this slope often becomes so steep that the mud slumps and slides down slope to be re-deposited at its foot. Slumping often appears to generate and move down gullies in the delta slope. There is some evidence in the Mississippi and

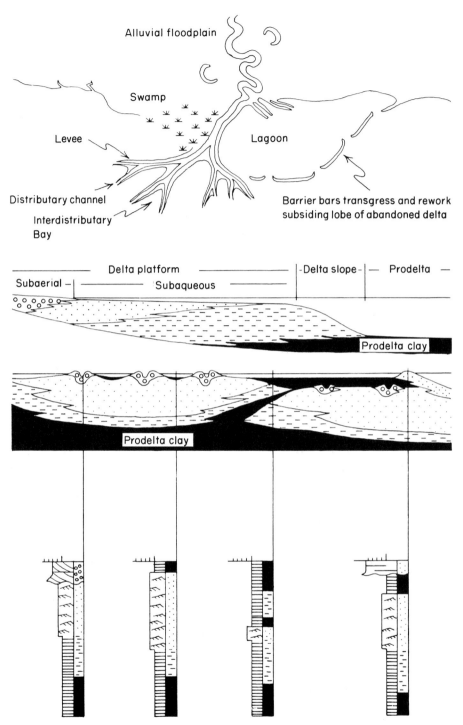

*Figure 5.1* Illustrative of the geomorphology and sedimentary facies of a Recent delta. Note the complexity of the vertical sections which may result.

Fraser deltas that this is associated with turbidity currents (Shepard, 1963, pp. 494 and 500).

Thus a vertical section of a delta reveals a fine-grained marine facies, which passes up transitionally into coarser freshwater sediments; these facies boundaries will be diachronous seaward as the delta builds out. Vertical repetition of this sequence may be expected, due to repeated crevassing and delta switching.

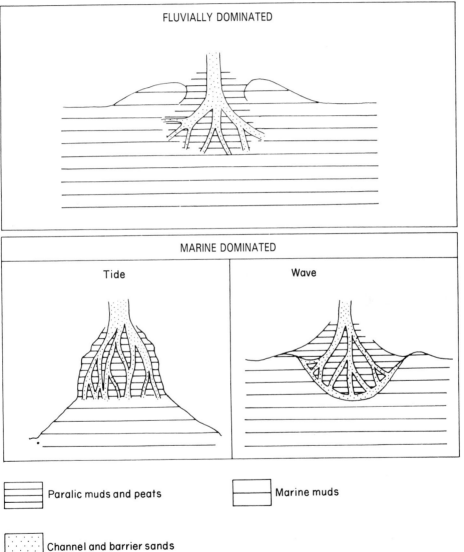

*Figure 5.2*  Sketch maps to illustrate the basic types of delta. For examples see text. Note how process controls the distribution and trend of sand bodies which may be potential hydrocarbon reservoirs.

In searching for ancient deltas, therefore, we must look for thick clastic sequences showing repeated cycles of upward-coarsening grain size. Each cycle should begin, at the base, with a marine shale which passed up through silts into coarser freshwater channel sands at the top. In plan the channels should show a radiating shoestring pattern and be cut into freshwater shales and coals. Coals may also be present on the top of the channels (Fig. 5.1).

Examination of modern deltas shows that they are in fact rather more complex than this simple picture. Two major types of delta may be differentiated: fluvially dominated deltas, and those dominated by marine process. The Mississippi is the classic example of the first type. It is in fact a very unusual delta with a very low ratio of sand to mud (1:9), and unusual in that it debouches into a sheltered embayment with a low tidal range. Thus distributary channels can radiate far out to sea. Deltas of this fluvially dominated 'birdfoot' pattern are uncommon except in lakes.

Most deltas are subjected to extensive marine reworking. An important distinction must be made between deltas dominated by wave or tidal action. The Nile and the Niger are examples of the first type. In these deltas sand is no sooner deposited at the mouth of a distributary than it is reworked by wave action and redeposited in an arc of barrier islands around the delta periphery.

Deltas in areas of high tidal range however have a quite different geomorphology. These have wide expanses of tidal flat fine sands and muds, often colonized by mangrove swamps. These are crosscut by braided distributary channels scoured by tidal currents. Instead of radiating from a point these channels tend to be subparallel. Tidally dominated deltas of this type are widespread in South-East Asia; examples include the Ganges: Brahmaputra, the Klang, and the Mekong.

The differentiation between these three types of delta is important in subsurface facies analysis because of the different distribution and orientation of potential hydrocarbon reservoirs (Fig. 5.2).

Ancient deltaic deposits are worldwide in their distribution and are known throughout geologic time.

The Carboniferous Period seems to have been a particularly favourable time for delta-formation and the following case history is of this age.

# CARBONIFEROUS DELTAIC SEDIMENTATION OF NORTHERN ENGLAND

At the end of the Devonian Period, a marine transgression took place across northern England. This sea advanced over a tectonically unstable region of actively moving fault-bounded blocks and basins. The sedimentary facies of the Lower Carboniferous are diverse, and their distribution is tectonically controlled. Due to years of study of their fauna, the complex stratigraphy of these rocks is largely resolved. Useful accounts of this work have been given by Rayner (1968) and Anderton *et al.*, (1979).

In general, it is possible to recognize five major sedimentary facies in the Dinantian and Namurian rocks of northern England:

(*i*) *Limestones*, including calcilutites, oolites, and bioclastic calcarenites with marginal reef complexes, were deposited on the relatively stable shelves.

(*ii*) *Black Shales* were deposited mainly in the basinal areas.

(*iii*) *The Millstone Grit* facies of coarse pebbly sandstones overlies parts of the shelf limestones and basinal shales.

(*iv*) *Greywacke sandstone and shale* facies, including the Mam Tor Series and Shale Grit of the Central Pennine Trough.

(*v*) *Yoredale* facies: cyclic sequences of limestone, shale, sandstone, and coal occur on the shelf areas above, or time equivalent to the limestone facies previously described.

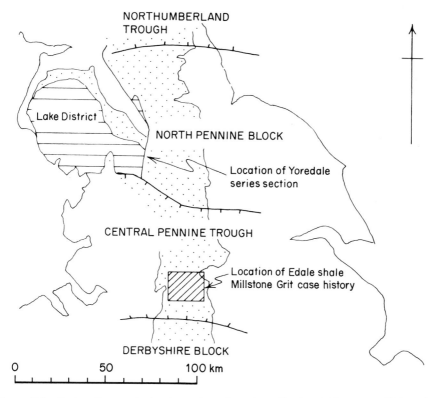

*Figure 5.3* Carboniferous tectonic setting of northern England. Outcrop of Dinantian and Namurian rocks stippled; older rocks horizontally ruled; younger rocks blank.

The last three of these five facies are believed to be essentially of deltaic origin. Two quite different types of delta seem to have operated. Their deposits will now be described and discussed under the arbitrary designations of Type 1 and Type 2, respectively.

## Type 1: description

The first type of Carboniferous deltaic sediments to be described is the Yoredale Series. These range from Dinantian to Namurian in age and extend from the Midland Valley of Scotland to the southern edge of the North Pennine Block (Fig. 5.3).

Basically the Yoredale Series consist of eight major cyclothems, each about 30 m thick, with the general motif: limestone, shale, sandstone, coal. Within

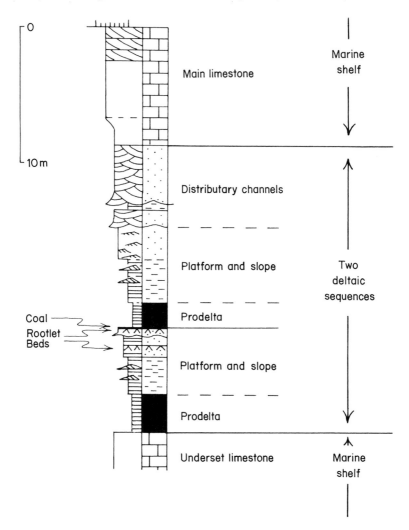

*Figure 5.4* Measured section of Yoredale Series, Hell Gill beck, Westmorland (for location see Fig. 5.3). Note how the overall upward-coarsening clastic sequence between the two limestones contains two subsequences, only the lower one of which is capped by a coal.

this main pattern minor cycles are present, notably thin shale, sandstone, coal sequences towards the top of major sandstones. Figure 5.4 shows two upward-coarsening shale, sandstone sequences between two limestones from the Yoredale Series of Westmorland.

The limestones of the Yoredale Series have extensive sheet geometries which are relatively easy to correlate laterally in contrast to the terrigenous beds. They are composed of a variety of carbonate rock types ranging from cross-bedded skeletal calcarenites to massive or thinly bedded calcilutites. Silicified limestones and thin chert layers occur especially in the upper part. Cross-bedding and terrigenous sand grains are often typical of the low part of each limestone sequence. Banks of crinoid ossicles and patch reefs of alga, bryozoa, corals, and brachiopods are not uncommon.

The limestones, within each cycle, are generally overlain by shales. These are black pyritic micaceous siltstones with thin siderite concretionary layers. They are laminated throughout and generally unfossiliferous.

The shales pass up gradually with increasing grain size into silty medium- to fine-grained sandstones which are generally laminated or micro-crosslaminated and sometimes burrowed. This sand facies is overlain abruptly by coarse cross-bedded sandstones. These occupy large channels, generally of only local extent but sometimes tens of feet deep. These often cut not only into the underlying sandstone, but are also sometimes incised into the shale, or sometimes even the limestone beneath (see for example Moore, 1959, Fig. 8).

The channel sands become finer grained towards the top where they are interbedded with thin coals, carbonaceous shales, and rootlet beds.

## Type 1: interpretation

The Yoredale Series consist of a series of cycles, each of which shows a progressive change from a marine environment (shown by the limestone fauna) to an upward-coarsening clastic sequence with rootlets at the top which indicate continental conditions. As discussed at the beginning of the chapter these are the features to be expected in a delta.

Considered in more detail the fauna and lithology of the limestones suggest a marine shelf environment with no influx of terrigenous sediment. Lime mud was deposited under low energy conditions, presumably below wave base. The cross-bedded skeletal calcarenites indicate intermittent current action (see Chapter 8 for further data on marine shelf environments). These conditions were terminated by an influx of terrigenous sediment. Initially, as the laminated shales show, this was fine grained and deposited in quiet relatively deep water out of suspension. The vertical increase in grain size and cross-lamination suggest progressive shallowing and increase in current action. Still stronger currents are indicated by the cross-bedded coarse channel sands. Once these were deposited, however, quiet conditions prevailed with peat swamps colonizing the abandoned channel surfaces and inter-channel areas.

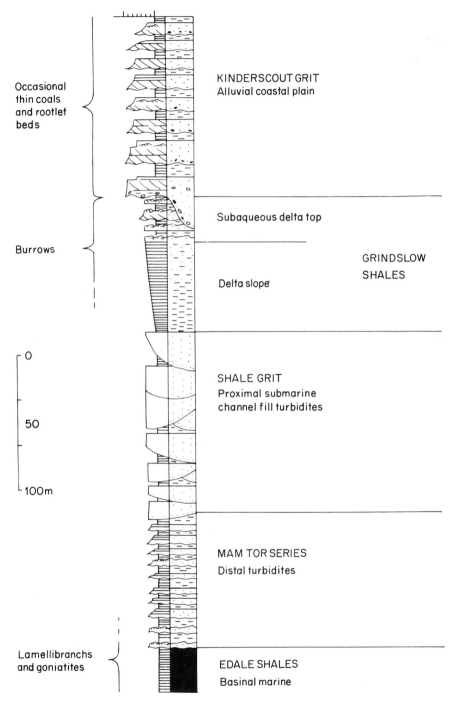

*Figure 5.5* Generalized composite section of the Edale Shale: Kinderscout Grit sequence. Location shown in Fig. 5.3. Based on data due to Collinson, Reading, and Walker (see references).

Considered in vertical sequence therefore the facies of the Yoredale Series can be attributed to the following environments:

| FACIES | ENVIRONMENT | |
|---|---|---|
| Coals and rootlet beds | Swamp | |
| Cross-bedded sands | Alluvial Distributary Channels | Deltaic |
| Rippled sands | Sub-aqueous delta platform | |
| Laminated silts | Prodelta | |
| Limestones | Marine shelf | |

This type of deltaic deposit which built out into open marine carbonate shelf facies is in marked contrast to that which formed where the deltas debouched into the deeper Carboniferous basins.

## Type 2: description

The second case history of deltaic sediments to be described from the Carboniferous of Northern England concerns the Millstone Grit and underlying sands and shales. An extremely well-documented example of this sequence is known from the northern edge of the Central Pennine trough. A condensed account of this work follows (see Figs 5.3 and 5.5).

The sequence begins, at the base, with Edale Shales which are about 200 m thick. These are dark grey mudstones often carbonaceous with disseminated pyrite. The only sedimentary structure which they show is a well-developed and laterally continuous lamination. The shales are largely unfossiliferous but occasionally yield thin shelled bivalves termed *Posidonia* and goniatites. The cephalopods are generally restricted to thin black horizons and are important biostratigraphic index fossils.

The Edale Shales pass up into the Mam Tor Series. These are about 120 m thick, consisting of laterally persistent interbedded sandstones and shales generally less than 1 m thick. The sandstones are fine-grained greywackes composed of grains of quartz, lithic fragments, and feldspars set in a carbonaceous clay matrix. The base of each sandstone is erosional, with flute and groove marks incised into the shale beneath. Internally the sands show graded bedding and a tendency for each unit to show the following structures arranged from bottom to top in the sequence: massive, flat-bedded, rippled.

The Mam Tor Series pass up into the Shale Grit, which has a thickness of between 100–200 m. This consists largely of greywacke sandstones petrographically similar to those beneath. Shales are rare. The sandstones are thicker bedded however than those of the Mam Tor Series, sometimes with sequences up to 30 m thick composed of sand beds 1–3 m thick. Graded bedding, lamination, and ripple marking are rare. Many of the sands are massive. These sandstones and the rarer shales are cut into and infill a number of spectacular channels up to 20 or more metres deep. These are most common in the upper part of the Shale Grit and also occur in the lower part of the overlying Grindslow Shales. These are about 100 m thick and succeed the Shale Grit with an abrupt base. They are sub-divided into an upper and a lower unit. Their lower part is an upward-coarsening sequence of laminated mudstones, siltstones, and fine flat-bedded and rippled silty sandstones. Burrows become increasingly common towards the top. The upper part of the Grindslow Shales consists of a heterogenous assemblage of siltstones and fine sandstones which are often laminated, rippled, and intensively burrowed. These are pierced by channels of coarser cross-bedded sand.

The Grindslow Shales are overlain by the Kinderscout Grit, the local representative of the Millstone Grit. This consists essentially of over 100 m of coarse pebbly arkosic sandstones with smaller quantities of shale.

At its base the Kinderscout Grit infills a series of spectacular channels incised

*Figure 5.6*   Tabular planar cross-bedding in coarse fluviatile sandstone. Kinderscout Grit. 'The Warren', Hathersage. Photo by courtesy of J. D. Collinson.

into the Grindslow Shales beneath. These are often over 30 m deep and nearly 300 m wide. They are infilled with very coarse pebbly sandstone, generally massive at the base and flat-bedded or cross-bedded higher up. The overlying part of the Kinderscout Grit consists of fining-upward sequences. These begin, at the base, with an erosion surface overlain by coarse cross-bedded sandstones often with quartz pebbles and shale fragments (Fig. 5.6). The sands become finer grained, laminated, and rippled upwards passing into the laminated grey siltstones with thin coals and rootlet horizons.

## Type 2: interpretation

The sequence from the Edale Shales to the Kinderscout Grit shows an overall vertical increase in grain size and a change from marine to continental conditions as shown by the goniatites and rootlet horizons respectively. These two criteria are typically deltaic as discussed at the beginning of the chapter. Considered in more detail, however, there are significant differences between the Edale-Kinderscout sequence and the Recent delta model and the Yoredale Series. In particular the Edale-Kinderscout sequence consists actually of two upward-coarsening sequences. These are of dissimilar facies and cannot be the result of delta switching.

The Edale Shales indicate deposition of mud out of suspension in a low energy environment. The goniatites show that this was marine. The disseminated pyrite indicates anaerobic stagnant reducing conditions. The graded greywackes of the Mam Tor Sandstones and the Grindslow Grit suggest deposition from turbidity currents. The reasons for this diagnosis are not given here. They are discussed in Chapter 10. The deep channels in the upper part of the Shale Grit were probably cut by the turbidity flows which debouched on to the basin floor to deposit the thinner more regularly bedded Mam Tor greywackes. It is possible therefore to designate the Grindslow Grit deposits 'proximal' turbidites which cut and infilled submarine canyons. The Mam Tor sandstones may be termed 'distal' turbidites deposited on fans radiating from the canyon mouths. This association of canyons and turbidite fans is well known from Recent nearshore marine basins such as the San Diego Trough (Hand and Emery, 1964).

The top of the Shale Grit marks the end of the first upward-coarsening sequence. The base of the overlying Grindslow Shales marks the start of the second.

The laminated silts and silty fine sands at the base of the Grindslow Shales suggest deposition out of suspension below wave base. The vertical transition of these deposits into the coarser rippled burrowed and channelled upper part of the Grindslow Shales suggest a progressive shallowing of the environment with deposition from traction currents becoming more and more important.

The deep channels at the base of the overlying Kinderscout Grit clearly herald the advance of an extremely violent current system with coarse sediment. The overlying upward-fining sand: shale cycles with coarse sediment. The overlying upward-fining sand: shale cycles suggest deposition from smaller channels which migrated laterally across an alluvial floodplain (p. 42). Coals and rootlets indicate intermittent sub-aerial exposure. Considered together the lower part of the Grindslow Shales can be attributed to deposition on a delta slope. The rippled burrowed silty sands above represent delta platform deposits laid down in interdistributary areas between the cross-bedded sandstone channels. Finally the delta top was buried by the advancing alluvial floodplain.

The Edale Shale-Kinderscout Grit sequence records therefore the gradual infilling of the Central Pennine trough first by turbidites and then by a delta. Palaeocurrent studies show that these deposits came into the area from the North Pennine block, the various facies presumably migrating diachronously southwards (Fig. 5.7). This sequence provides a classic demonstration of Walther's Law (see p. 3); a vertical conformable sequence of facies produced by the lateral migration of a series of environments.

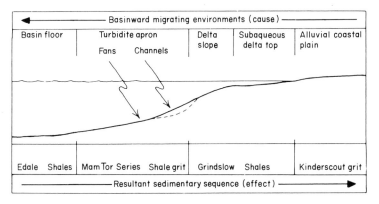

*Figure 5.7* Cross-section diagram illustrating the relationship between facies and environments in the Carboniferous rocks of the Central Pennine Trough.

## GENERAL DISCUSSION OF DELTAIC SEDIMENTATION

The two case histories just described illustrate two contrasting types of delta. The Yoredale deltaic deposits built out relatively thin clastic wedges on to shallow marine carbonate shelf deposits.

The Millstone Grit delta on the other hand built out in to a deep marine basin. A steeper delta slope developed therefore and generated aprons of turbidite deposits at its foot. There are other examples of ancient deltaic sediments which can be attributed to the high slope and low slope varieties just defined. High slope deltas, where marine shales pass up through turbidites to channel sands, have been described from the Coaledo Formation (Eocene) of Oregan (Dott, 1966), from Ordovician rocks of the Appalachians (Horowitz, 1966), and from the Westphalian of North Devon (De Raaf, Reading, and Walker, 1965). Low

*Figure 5.8* (opposite page) Composite measured section of coarsening-upward cycles from the Tabuk basin, Arabia.

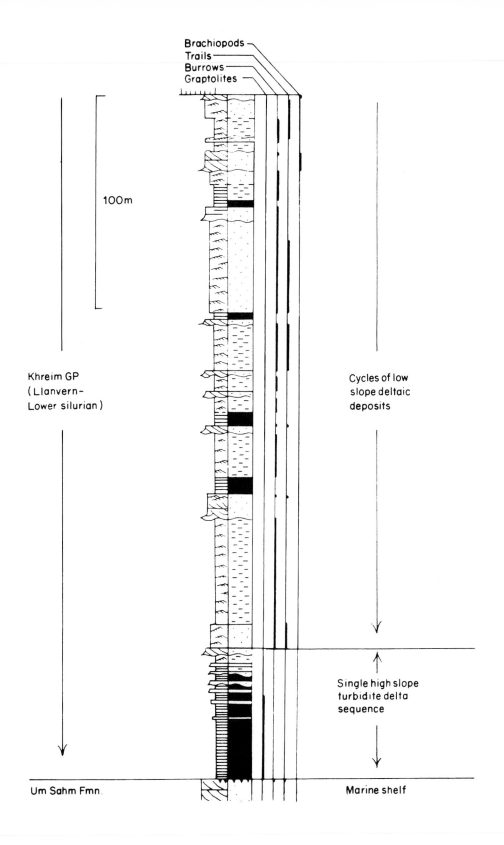

Brachiopods
Trails
Burrows
Graptolites

100m

Khreim GP
(Llanvern-
Lower silurian)

Cycles of low
slope deltaic
deposits

Single high slope
turbidite delta
sequence

Um Sahm Fmn.

Marine shelf

slope deltaic deposits where marine shales (and sometimes limestones) pass through a clean-washed sheet sand facies into alluvial channels are typical of the Pennsylvanian (Upper Carboniferous) of the Illinois basin, USA. The Anvil Rock sandstone is a particularly good example of this type. A lower sheet phase, some 7 m thick, is laminated or rippled and passes down transitionally into marine shale. Locally, the sheet phase is cut through by coarser cross-bedded channel sands which meander down the palaeoslope (Hopkins, 1958; Potter and Simons, 1961). The Khreim Group (Ordovician-Silurian) of the Tabuk basin of Arabia show a series of deltaic cycles, the lowest of which is a high slope turbidite variant. The overlying cycles suggest relatively low slope conditions. Within each sequence laminated silts pass up through rippled and burrowed interlaminated silts and sands, which underly the upper channel sandstones (Fig. 5.8).

It might be thought from the preceding discussion that such well-documented deposits as deltas were fully understood. This is not true. Many problems remain to be solved, particularly the origin of cyclicity in deltas. This phenomenon is common in ancient deltaic deposits. It has been studied intensively, especially in Carboniferous rocks in which the distribution of deltaic facies seems to be nearly world-wide (Duff, Hallam, and Walton, 1967, p. 81). Studies of Recent deltas suggest that the crevassing of distributary channels provides a built-in cycle generator, no external cause need be invoked (Moore, 1959; Coleman and Gagliano, 1964). Explanations of deltaic cycles which ignore this fact seem improbable (e.g. the tectonic hiccup hypothesis for the Yoredale cycles proposed by Bott and Johnson in 1967). On the other hand, not all cycles in deltaic sediments can be attributed to delta switching. It is possible to correlate individual marine marker bands from Carboniferous deltas as far apart as Ireland and Germany (Ramsbottom and Calver, 1962). Since these extend beyond individual deltas and basins, they can be due neither to crevassing nor sporadic basin subsidence. Eustatic changes are a much more probable explanation for this phenomenon (e.g. Wanless and Shepard, 1936).

It would seem, therefore, that, while deltas can generate their own cycles, external factors such as regular tectonic bumps, climatic and sea-level changes, will also leave their marks on the deltaic pile. In understanding deltaic sequences it is necessary to develop criteria to distinguish the local delta cyclic motif (which must be present) from cycles of external origin which may or may not be present.

There are two useful lines of approach which may help to solve this problem; both employ computers to perform complex statistical gymnastics. The first approach is to study the thickness and regional variation of cyclothems and to see how these relate to other geological variables (e.g. Read and Dean, 1967; 1968).

The second approach uses the computer to simulate deltaic sedimentation; calibration of variables such as sediment input and rate of subsidence being based on Recent examples. (Oertel and Walton, 1967; Bonham-Carter and Sutherland, 1967; Bonham-Carter and Harbaugh, 1968.)

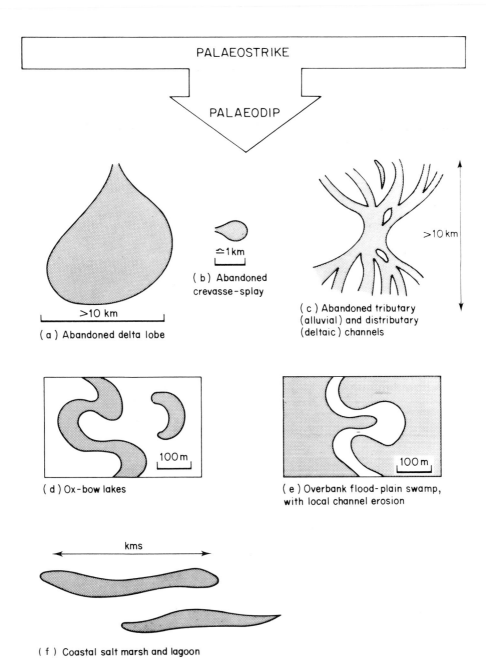

PALAEOSTRIKE

PALAEODIP

(a) Abandoned delta lobe
>10 km

(b) Abandoned crevasse-splay
≃1km

(c) Abandoned tributary (alluvial) and distributary (deltaic) channels
>10 km

(d) Ox-bow lakes
100 m

(e) Overbank flood-plain swamp, with local channel erosion
100 m

(f) Coastal salt marsh and lagoon
kms

*Figure 5.9* Cartoons to illustrate some of the ways in which coal beds may occur. Note that geometry is a function both of depositional environment and the effects of subsequent erosion.

## ECONOMIC SIGNIFICANCE OF DELTAIC DEPOSITS

Deltaic sediments are important sources of coal, oil, and gas. Peat-formation is typical in the swamps and marshes of Recent deltaic alluvial plains. Detailed accounts of the sedimentology of coal deposits are given by Dapples and Hopkins (1969), Klein (1980), Ward (1983), and Galloway and Hobday (1983). In the effective exploitation of coals, it is obviously important to understand their depositional environment. Not all coals are deltaic; as already mentioned some occur in continental basins far from the sea (p. 71). Within the broad framework of a delta, coals can form in a number of different situations.

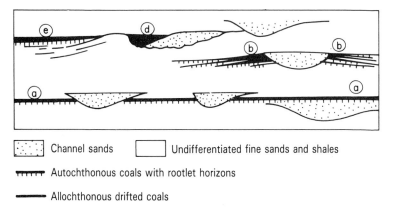

▱ Channel sands   ▢ Undifferentiated fine sands and shales

ⲧⲧⲧⲧⲧ Autochthonous coals with rootlet horizons

▬▬ Allochthonous drifted coals

*Figure 5.10* Cross-section to show the lateral continuity and setting of different types of coal bed. The section is orientated parallel to the palaeostrike. Letters correlate with coal geometries illustrated in Fig. 5.9.

There are many different environments in which swamps grow. Therefore there are many different ways in which ancient coal beds may occur. Wanless, Barroffio and Trescott (1969) published a classic account of the relationship between depositional environments and coal bed geometry in the Pennsylvanian deltas of Illinois. Other examples will be found in Klein (1980) and Galloway and Hobday (1983). Figures 5.9 and 5.10 illustrate some of the different types of coal in plan and cross-section respectively.

Sometimes when a major delta is abandoned or when there is a major eustatic rise in sea level blanket deposits of peat may form (Fig. 5.9 a). On a smaller scale individual crevasse-splays may be covered by a bed of peat (Fig. 5.9 b). Repetition of this process can deposit coals that thin and split away from the channel margin (Fig. 5.10 b).

Peat formation can also take place in abandoned fluvial tributary and deltaic distributary channels (Fig. 5.9 c); a minor variety of this situation is to be found in oxbow lakes (Fig. 5.9 d). Blanket peats may form in flood basins on alluvial

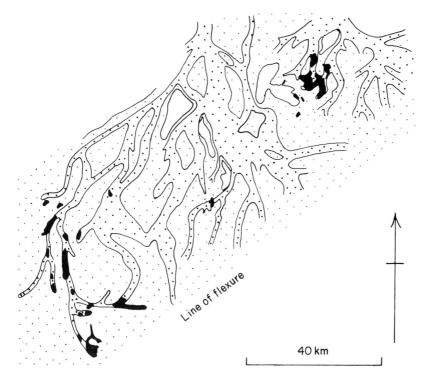

*Figure 5.11* Isopach map of the Booch Sandstone (Pennsylvanian) of Oklahoma. Sheet phase (0–20 ft) sparse stipple, channel phase (20–200 ft) dense stipple. See how oil fields (black) are concentrated in the channel sands. Modified from Busch, 1961, Fig. 12. Reproduced by courtesy of the American Association of Petroleum Geologists.

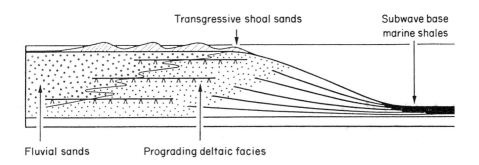

*Figure 5.12* Geophantasmogram to show the relationship between the three facies of the Jurássic reservoir sands of the North Sea. From Selley, 1976a, by courtesy of the Geologists' Association.

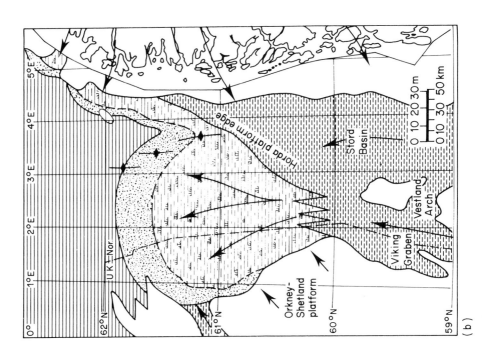

(a)

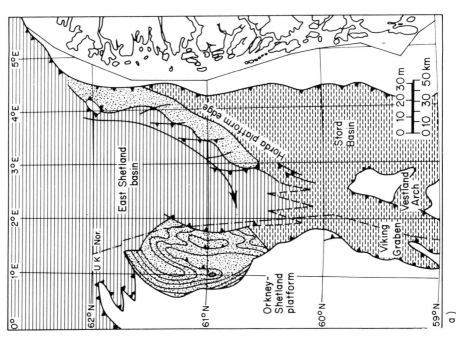

(b)

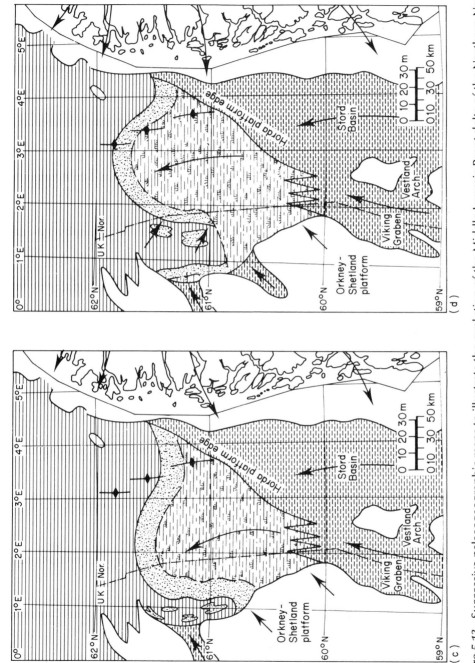

*Figure 5.13* Successive palaeogeographic maps to illustrate the evolution of the Middle Jurassic Brent delta of the Northern North Sea. From Eynon, 1981, by courtesy of the Institute of Petroleum.

plains (Fig. 5.9 e). Note that blanket coals can be locally eroded by channels (Fig. 5.10 a). In such areas it is obviously very important to be able to map and predict such channel trends. This is not just because sandstone channels cut out valuable coal reserves, but also because they are sometimes highly permeable aquifers that can cause the flooding of underground mine workings. Lastly coal can occur in elongate bodies parallel to the local shoreline (Fig. 5.9 f). This is where peat grew in salt marsh and coastal lagoon environments.

This brief review of the relationship between coal bed geometry and depositional environment shows that facies analysis can aid the exploitation of coal deposits.

Deltas are often major hydrocarbon provinces for a number of reasons. The deltaic process is an excellent means for transporting sands (potential hydrocarbon reservoirs) far out into marine basins with organic rich muds (potential hydrocarbon source beds).

Because deltas often form in areas of crustal instability structural deformation forms traps for migrating hydrocarbons. Deltas also form their own traps in growth faults, roll-over anticlines and associated stratigraphic traps. (Fisher *et al.*, 1972; Selley, 1977; Coleman 1982).

Obviously it is important to understand the sedimentology of deltas so as to predict the location and geometry of a hydrocarbon reservoir. Figure 5.11 illustrates a classic example of petroleum stratigraphically trapped within a system of deltaic distributaries. Some of the modes of channel entrapment discussed for alluvial channels also apply to deltaic ones (refer back to p. 69 and Fig. 2.18). Even where oil is structurally trapped in deltaic reservoirs environmental interpretation has an important part to play in both petroleum exploration and development.

Many oil fields in the northern North sea produce from tilted truncated fault block traps. The main reservoir is the Middle Jurassic Brent Sandstone. This was deposited by a delta system that prograded northwards down the axis of the Viking Graben. Three major facies can be recognized: coarse pebbly non-marine sands and shales (fluvial), carbonaceous shales, sands and coals (deltaic), and clean sands that are occasionally shelly, glauconitic and burrowed (marine shoal). These are arranged in regressive: transgressive units (Fig. 5.12). There is a close correlation between facies and reservoir quality (porosity and permeability). Careful facies analysis has made it possible to map the progradation and destruction of the Brent delta. Palaeogeographic maps may thus be used to predict regional variations in reservoir quality (Fig. 5.13). Figure 5.14 illustrates the stratigraphy of the Brent sequence. The lower progradational phase of the delta deposited units with considerable lateral continuity. The upper part of the sequence is much more varied with local phases of transgression and truncation. During the development of fields there have been problems in differentiating deltaic channel sands from marine ones. The former are locally aligned down the delta slope, while the latter tend to parallel it (Fig. 5.15).

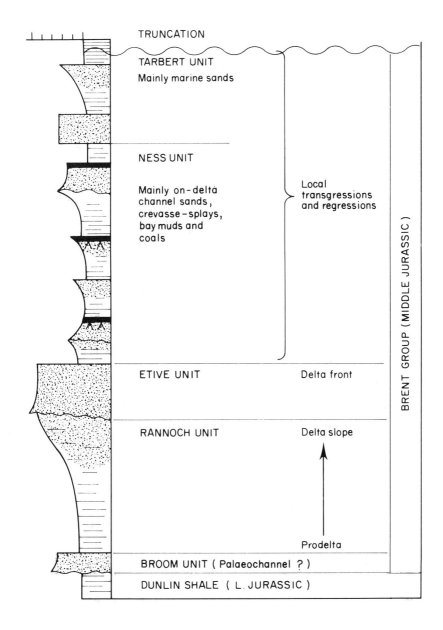

*Figure 5.14* Summary stratigraphic section of the Middle Jurassic Brent Group of the Northern North Sea. Reservoir continuity is regular in the lower part, but becomes more complex towards the top due to the progradation and retreat of local subdelta lobes with transgressive marine sands.

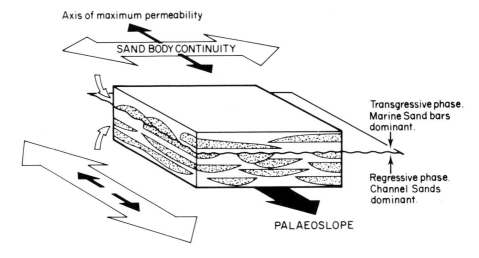

*Figure 5.15*   Cartoon to illustrate the relationship between reservoir continuity in a heterogenous delta complex with interbedded delta channel and marine shoreface sands.

## SUB-SURFACE DIAGNOSIS OF DELTAIC DEPOSITS

Deltaic deposits may be easily recognized in the sub-surface. The geometry of a delta may often be recognized before a single well has been drilled. Major progradational foreset bedding is commonly seen in seismic sections through deltaic sediments (Fig. 5.16). On the top of the delta individual channels may be mappable, as discussed when fluvial channels were dealt with (p. 74). Coals have anomalously low seismic velocities so they may be identifiable when they are thick or shallow. Once wells have been drilled it may be possible to map individual delta lobes from palaeontology and wireline log correlations.

Lithologically deltas are characterized by major wedges of terrigenous clastics with widely varying grain-size and sand:shale ratios. Coal or lignite beds are common and disseminated carbonaceous detritus occurs in distributary channel sands, and other on-delta facies. In the shoreface sands of wave dominated deltas, by contrast, carbonaceous detritus may be absent, and glauconite may occur instead.

A major deltaic lobe will show an overall upward-sanding sequence on wireline logs. In detail however all four major log motifs may be found. Figure 5.17 shows the sequence of log motifs and sedimentary structures which may be anticipated in deltaic deposits. Weber (1971) has given a masterly demonstration of how detailed log interpretation can unravel the complex deltaic sands and shales of the Niger delta. Figure 5.18 shows a borehole through deltaic sediments of the North Sea.

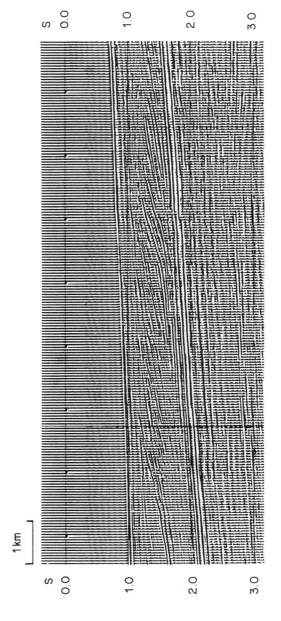

*Figure 5.16* Seismic section through a delta showing progradation from topset through forset to bottomset. From Fitch, 1976. Fig. 26. By courtesy of Gebruder Borntraeger.

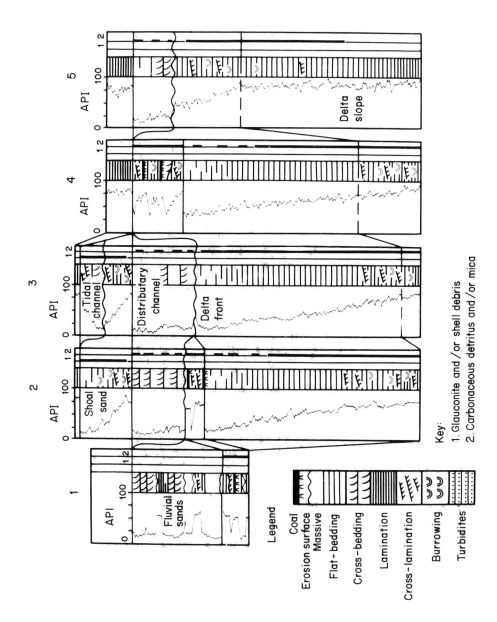

Legend

Coal
Erosion surface
Massive
Flat - bedding
Cross - bedding
Lamination
Cross - lamination
Burrowing
Turbidites

Key:
1. Glauconite and / or shell debris
2. Carbonaceous detritus and / or mica

1
API  0  100  1 2
Fluvial
sands

2
API  0  100  1 2
Shoal
sand

3
API  0  100  1 2
Tidal
channel
Distributary
channel
Delta
front

4
API  0  100  1 2

5
API  0  100  1 2
Delta
slope

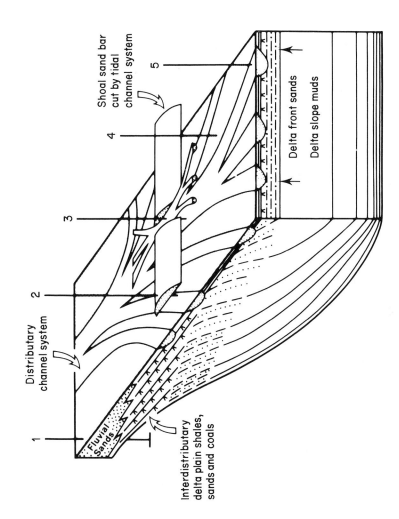

*Figure 5.17* Cartoon to illustrate the sand bodies and log characteristics of deltaic sedimentary facies. When core material is available suites of sedimentary structures are often diagnostic. Even without core material, however, depositional environment may be detected from log motif (grain-size profile) and the distribution of glauconite and carbonaceous detritus.

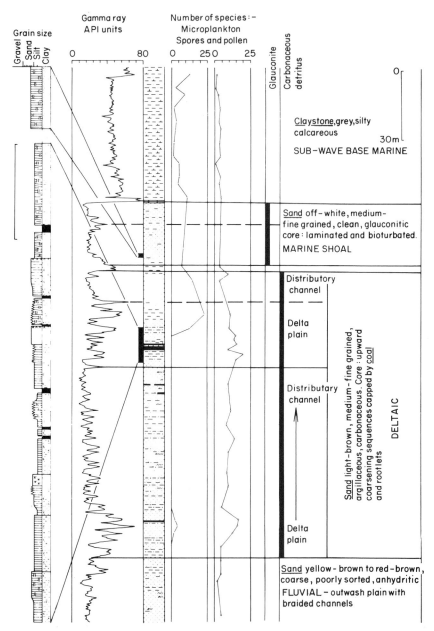

*Figure 5.18* A borehole through deltaic deposits beneath the North Sea. This shows how red sands and shales were deposited in a fluvial environment. These are succeeded by carbonaceous sands and shales with thin coals. Both the cores and the gamma log show upward-sanding pro-gradational sequences.

A subsequent marine transgression deposited glauconitic shoal sands on the abandoned delta. These were buried in turn beneath marine muds. From Selley (1976). Fig. 4, by courtesy of the *American Association of Petroleum Geologists*.

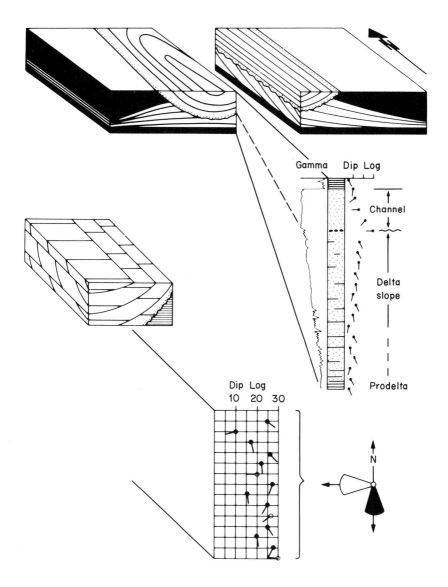

*Figure 5.19* Illustration to show the dip motifs found in deltaic deposits. Upward-increasing blue motif in the delta slope shows that this example prograded to the south. The upward-decreasing red pattern in the distributary channel reflects major bedding surfaces striking north-south along channel axis and dipping west into it.

Close spaced dips (below) may indicate that the red readings dip west (open tadpoles); while steeper dips are due to southerly cross-beds dipping down channel (black tadpoles). From Selley (1977) reproduced by courtesy of *Applied Science Publications*.

Similarly dipmeter patterns in deltas are often complex but, if correctly interpreted, can make a valuable contribution to the unravelling of facies geometries (Gilreath and Stephens, 1971).

Two major types of motif are generally present. Upward-increasing 'blue' patterns characterize progradational upward-sanding sequences. This will dip in the direction of outbuilding. Plotting back of such motifs in one delta lobe may thus locate the sandiest part, where the optimum reservoir conditions may be found.

Delta distributary channels may show very complex dip patterns. When the dips from one channel are plotted on a polar plot, however, a 90° bimodal pattern can appear. This may show one mode with high angle cross-bedding which dips down the channel axis, and another mode which forms a red (upward-declining) pattern on the log. This reflects major point-bar foresets which dip into the channel axis (Fig. 5.19).

This type of analysis of channels, which is generally applicable to fluvial channels, is extremely important. From a single well it is possible to ascertain both the channel trend, and the direction in which the axis lies.

Information of this type is invaluable when looking for stratigraphically trapped oil in channel sands, as well as in regional palaeogeographic studies.

## REFERENCES

A very brief list of works on Recent deltas, and reviews of Recent and ancient deltaic sedimentation:

Allen, J. R. L., 1965. Late Quarternary Niger Delta, and adjacent areas: Sedimentary environments and lithofacies. *Bull. Amer. Assoc. Petrol. Geol.*, **49**, 547–600.

Coleman, J. M., 1982. *Deltas: Processes of Deposition and Models for Exploration.* I. H. R. D. C., Boston. p. 124.

Coleman, J. M., and Wright, L. D., 1975. Modern river deltas; variability of processes and sand bodies. In: *Deltas: models for exploration.* (Ed. Broussard, M. L.) pp. 99–149.

Coleman, J. M., and Prior, D. B. 1982. Deltaic Environments. In: Sandstone Depositional Environments. P. A. Scholle and Spearing, D. (Eds.). Amer. Assoc. Petrol. Geol. Tulsa. 139–178.

Fisk, H. N., 1961. Bar finger sands of Mississippi Delta. In: *Geometry of Sandstone Bodies.* (Ed. J. A. Peterson and J. C. Osmond) Amer. Assoc. Petrol. Geol. pp. 29–52.

——, McFarlan, E., Kolb, C. R., and Wilbert, L. J., 1954. Sedimentary Framework of the modern Mississippi Delta. *J. Sediment. Petrol.*, **24**, 76–100.

Galloway, W. E. and Hobday, D. K. 1983. *Terrigenous Clastic Depositional Systems.* Springer-Verlag, Berlin. p. 421.

Klein, G. de V. 1980. *Sandstone Depositional Models for Exploration for Fossil Fuels.* CEPCO, Minneapolis. p. 149.

Moore, D., 1966. Deltaic Sedimentation. *Earth Science Reviews*, **1**, 87–104.

Moore, G. T., and Asquith, D. O., 1971. Delta: term and concept *Bull. Geol. Soc. Amer.*, **82**, 2563–8.

Morgan, J. P., 1970. Depositional processes and products in the deltaic environment. In: Deltaic sedimentation ancient and modern. *Soc. Econ. Pal. Min.* Sp. Pub. No. 15. (Eds J. P. Morgan and R. H. Shaver) pp. 31–47.

——, and Shaver, R. H., 1970. Deltaic sedimentation ancient and modern. *Soc. Econ. Pal. Min.* Sp. Pub. No. 15. 312–p.

Shepard, F. P., Phleger, F. B., and Van Andel, T. H. (Eds), 1960. *Recent Sediments Northwest Gulf of Mexico.* Amer. Assoc. Petrol. Geol. 394 p.

Shirley, M. L., and Ragsdale, J. A. (Eds), *Deltas in their Geologic Framework.* Houston. Geol. Soc. 251 p.

Oomkens, E., 1967. Depositional sequences and sand distribution in a deltaic complex. *Geol. Mijnb.*, **46**, 265–78.

Van Andel, T. H., 1967. The Orinoco Delta. *J. Sediment Petrol.*, **37**, 297–310.

Van Straaten, L. M. J. U., 1960. Some recent advances in deltaic sedimentation. *Lpool. Manchr. Geol. J.*, **2**, 441–43.

Wright, L. D., and Coleman, J. M., 1973. Variations in morphology of major river deltas as functions of ocean wave and river discharge regimes. *Bull. Amer. Assoc. Petrol. Geol.*, **57**, 370–417.

The description of Carboniferous deltaic sedimentation of England is based on:

Allen, J. R. L., 1960. The Mam Tor Sandstones: A 'turbidite' facies of the Namurian deltas of Derbyshire, England. *J. Sediment. Petrol.*, **30**, 193–208.

Anderton, R., Bridges, P. H., Leeder, M. R. and Sellwood, B. W. 1979. *A Dynamic Stratigraphy of the British Isles.* George Allen & Unwin, London. p. 301.

Bennison, G. M., and Wright, A. E., 1969. *The geological history of the British Isles.* Edward Arnold, London p. 406.

Collinson, J. D., 1966. Antidune bedding in the Namurian of Derbyshire, England. *Geol. Mijnb.*, **45**, 262–4.

——, 1969. The sedimentology of the Grindslow shales and the Kinderscout Grit: A Deltaic Complex in the Namurian of Northern England. *J. Sediment. Petrol.*, **39**, 194–221.

——, 1970. Deep channels, massive beds and turbidity current genesis in the central Pennine basin. *Proc. Yorks. Geol. Soc.*, **37**, 495–519.

Johnson, G. A. L., 1959. The Carboniferous Stratigraphy of The Roman Wall district in western Northumberland. *Proc. Yorks. Geol. Soc.*, **32**, 83–130.

Moore, D., 1958. Yoredale Series of Upper Wensleydale and adjacent parts of northwest Yorkshire. *Proc. Yorks. Geol. Soc.* **31**, 91–146.

——, 1959. Role of deltas in the formation of some British Lower Carboniferous cyclothems. *J. Geol.*, **67**, 522–39.

Rayner, D. H., 1967. *The Stratigraphy of the British Isles.* Cambridge University Press.

Reading, H. G., 1964. A review of the factors affecting the sedimentation of the Millstone Grit (Namurian) in the Central Pennines. In: *Deltaic and Shallow Marine Deposits.* (Ed. L. M. J. U. Van Straaten) Elsevier, Amsterdam, pp. 340–6.

Walker, R. G., 1966a. Shale Grit and Grindslow Shales; transition from turbidite to shallow water sediments in the Upper Carboniferous of northern England. *J. Sediment. Petrol.*, **36**, 90–114.

——, 1966b. Deep channels in turbidite-bearing formations. *Bull. Amer. Assoc. Petrol. Geol.*, **50**, 1899–917.

Bonham-Carter, G. F., and Harbaugh, J. W., 1968. Simulation of geologic systems: an overview. In: Colloquium on Simulation. (Ed. D. F. Merriam and N. C. Cocke) *Computer Contribution* No. 22. State Geol. Surv. Univ. Kansas. pp. 3–10.

Bonham-Carter, G. F., and Sutherland, A. J., 1967. Diffusion and settling of sediments at river mouths: a computer simulation model. *Trans. Gulf Coast Assoc. Geol. Soc.*, **17**, 326–38.

Bott, M. H. P., and Johnson, G. A. L., 1967. The controlling mechanism of Carboniferous cyclic sedimentation. *Quart. J. Geol. Soc. Lond.*, **122**, 421–41.

Busch, D. A., 1961. Prospecting for stratigraphic traps. In: *The Geometry of Sandstone Bodies*. (Ed. J. A. Peterson and J. C. Osmond) Amer. Assoc. Petrol. Geol. pp. 220–32.

Coleman, J. M., and Gagliano, S. M., 1964. Cyclic sedimentation in the Mississippi River deltaic plain. *Gulf Coast Assoc. Geol. Soc. Trans.*, **14**, 67–80.

Dapples. E. C., and Hopkins, M. E. (Eds), 1969. Environments of Coal Deposition. *Geol. Soc. Amer.*, Sp. Pap., No. 114.

De Raaf, J. F. M., Reading, H. G., and Walker, R. G., 1964. Cyclic sedimentation in the Lower Westphalian of north Devon. *Sedimentology*, **4**, 1–52.

Dott, R. H., 1966. Eocene Deltaic Sedimentation at Coos Bay, Oregon. *J. Geol.*, **74**, 373–420.

Duff, P. McL. D., Hallam, A., and Walton, E. K. 1967. *Cyclic Sedimentation*. Elsevier, Amsterdam p. 280.

Eynon, G., 1981. Basin development and sedimentation in the Middle Jurassic of the Northen North Sea. In: (Eds G. D. Hobson and L. V. Illing). Institute of Petroleum, London. 196–204.

Fisher, W. L., Brown, L. F., Scott A. J., and McGowen, 1972. *Delta systems in the exploration for oil and gas*. Bureau Econ. Geol. Texas Univ., p. 127.

Fitch, A. A., 1976. *Seismic Reflection Interpretation*. Gebruder Borntraeger, Stuttgart. p. 147.

Gilbreath, J. A., and Stephens, R. W., 1971. Distributary front deposits interpreted from dipmeter patterns. *Trans. Gulf Coast Assn. Geol. Socs.*, **21**, 233–43.

Hand, B. M., and Emery, K. O., 1964. Turbidites and topography of North end of San Diego Trough, California. *J. Geol.*, **72**, 526–42.

Herodotus, *c.* 430 B.C. *The Histories of Herodotus of Halicarnassus*. Papyrus Publishing Coy. Old Cairo, also 1962. Translated by H. Carter, Oxford Univ. Press. p. 617.

Hopkins, M. E., 1958. Geology and Petrology of the Anvil Rock Sandstone of Southern Illinois. *Ill. State. Geol. Surv. Circ.*, No. 256, p. 49.

Horowitz, D. H. 1966. Evidence for deltaic origin of an Upper Ordovician sequence in the Central Appalacians. In: *Deltas in their Geologic Framework*. (Ed. M. L. Shirley and J. A. Ragsdale) Houston. Geol. Soc. pp. 159–69.

Lyell, Sir Charles, 1854. *Principles of Geology*. 9th Edn. Appleton & Coy, New York.

Oertel, G., and Walton, E. K., 1967. Lessons from a feasibility study for computer models of coal-bearing deltas. *Sedimentology,* **9**, 157–68.

Potter, P. E., and Simons, J. A., 1961. Anvil rock sandstone and channel cutouts of Herrin (No. 6). Coal in West-Central Illinois. *Ill. State Geol. Surv. Circ.*, **34**, 12.

Ramsbottom, W. H. C., and Calver, M. A., 1962. *Comptes Rendus 4ᵉ. Congr. Advanc. Etud. Stratigr. Carb.*, pp. 571–6.

Read, W. A., and Dean, J. M., 1967. A quantitative study of a sequence of coal-bearing cycles in the Namurian of central Scotland, 1. *Sedimentology*, **9**, 137–56.

Read, W. A., and Dean, J. M., 1968. A quantitative study of a sequence of coal-bearing cycles in the Namurian of central Scotland, 2. *Sedimentology*, **10**, 121–36.

Selley, R. C., 1976a. The Habitat of North Sea Oil. *Proc. Geol. Ass. London*. **87**. 359–88.

——, 1976b. Sub-surface environmental analysis of North Sea sediments. *Bull. Amer. Assoc. Petrol. Geol.* **60**, 184–95.

——, 1977. Deltaic facies and petroleum geology. In: *Developments in petroleum geology*. (Ed. G. D. Hobson) Applied Science Publishers Ltd. pp. 197–224.

Shepard, F. P., 1963. Submarine Canyons. In: *The Sea*, vol. III. (Ed. M. N. Hill) Wiley, New York. pp. 480–506.

Wanless, H. R., and Shepard, F. P., 1936. Sea level and climatic changes related to Late Paleozoic cycles. *Bull. Geol. Soc. Amer.*, **47**, 1177–206.

——, Barroffio, J. R., and Trescott, P. C., 1969. Conditiions of deposition of Pennysylvanian coal beds. In: Environments of Coal Deposition. (Ed. E. C. Dapples and M. E. Hopkins) *Geol. Soc. Amer., Sp. Pap.*, No. 114, pp. 105–42.

Ward, C. R., 1983. *Coal Geology: Exploration, Mining, Preparation and Use*. Blackwells, Oxford. p. 300.

Weber, K. J., 1971. Sedimentological aspects of oil fields in the Niger Delta. *Geol. Mijnb.*, **50**, 559–76.

# 6     Linear terrigenous shorelines

## INTRODUCTION:
## RECENT LINEAR TERRIGENOUS SHORELINES

Deltas only form where rivers bring more sediment into the sea than can be re-worked by marine current. By their very nature, therefore, deltaic sequences indicate a regression of the shoreline. Where marine currents are strong enough to redistribute land-derived sediment, linear shorelines are formed with bars and beaches running parallel to the coast.

Both deltas and linear shorelines deposit sediment in a wide range of sedimentary environments ranging from continental to marine. Studies of Recent sediments show that both linear and lobate shorelines can form upward-coarsening regressive sequences. Because deltas and linear shorelines both deposit porous sands around marine basins they are important hydrocarbon reservoirs. Since their sand body geometries are quite different, however, it is important to be able to distinguish the two types.

Studies of Recent terrigenous linear shorelines suggest that four major sedimentary environments can be recognized (e.g. Guilcher, 1970; De Raaf and Boersma, 1971; Steers, 1971; King, 1972; Meyer, 1972; Hails and Carr, 1975; Ginsberg, 1975; Davis and Ethington, 1976; Hobday and Erikson, 1977; Davis, 1978; and Swift and Palmer, 1978). These consist of two high-energy zones which alternate seawards with two low-energy zones. Basically from land to sea these are: fluviatile coastal plain, lagoonal and tidal flat complex, barrier island, and offshore marine shelf (Fig. 6.1). Each of these four environments deposit facies which can be distinguished from one another by their lithology, sedimentary structures, and biota. These will now be summarized. Further data are given in the references previously mentioned.

The fluviatile coastal plain will deposit alluvium similar to that described in Chapter 2. Where the coastal plain has a gentle gradient the alluvium will be of the meandering river type. Where it is steep, then a braided outwash plain is generally developed. In the more normal former case fine-grained flood-plain sediments will dominate over upward-fining channel sand sequences. The biota is continental, with bones, wood, and other plant debris and freshwater invertebrates.

The alluvial coastal plain passes transitionally seawards into swamps, tidal flats, and lagoons. Peat forms in the swamps. Tidal flats deposit delicately interlaminated muds, silts, and very fine sands, often rippled and burrowed. They are cut by meandering gullies in which channel floor lag conglomerates

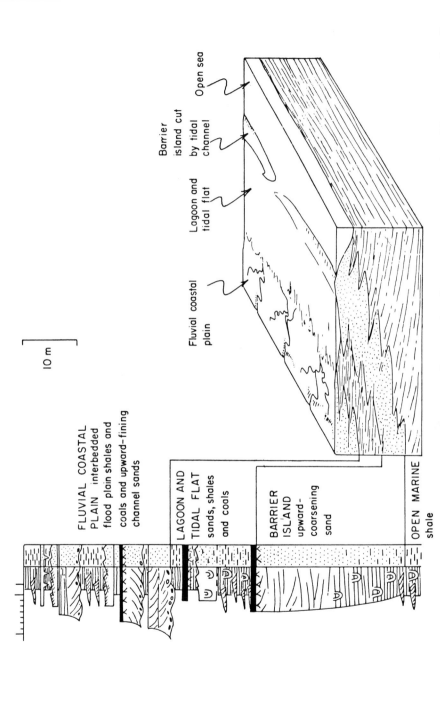

10 m

FLUVIAL COASTAL PLAIN interbedded flood plain shales and coals and upward-fining channel sands

LAGOON AND TIDAL FLAT sands, shales and coals

BARRIER ISLAND upward-coarsening sand

OPEN MARINE shale

Fluvial coastal plain

Lagoon and tidal flat

Barrier island cut by tidal channel

Open sea

*Figure 6.1* Geophantasmogram to show the environments, facies and sedimentary sequence produced by a prograding linear terrigenous coastline (From Selley, 1982, by courtesy of Academic Press).

are overlain obliquely by interlaminated fine sediment deposited on prograding point-bars (Van Straaten, 1959; Evans, 1965).

The deposits of lagoons are generally fine-grained too, but depending on their size and depth, may deposit sediment ranging from sand to mud. In regions of low sediment influx carbonate mud can be deposited. Evaporites may form in hypersaline lagoons. The fauna of lagoons is similarly variable depending on the salinity. It may range from freshwater, through brackish (with shell banks) to normal marine or, if restricted and of high salinity, a fauna may be absent. Sedimentary structures of lagoons are similar to those of tidal flats with delicately laminated muds and interlaminated and rippled sand silt and mud. Bioturbation is common.

The lagoon is separated from the open sea by a barrier island complex. This is composed dominantly of well-sorted sand with a fragmented derived marine fauna. A review of Recent barrier deposits (Davis 1978) suggests that the typical internal structure is regular bedding that dips gently seawards. Rare trough and tabular planar cross-bedding dips generally landwards, though bipolar onshore: offshore dips may be recorded (Klein, 1967). The barrier may be no more than an offshore bar exposed only at low tide, or it can form an island with eolian dunes on the crest. Intermittently along its length the barrier may be cut by tidal channels in which cross-bedded sands are deposited (Armstrong Price, 1961; Hoyt and Henry, 1967). To landward the barrier may pass abruptly into the lagoon with the development of washover fans. Alternatively a tidal flat may intervene. Barriers typically pass seawards with decreasing grain size into an offshore zone where mud is deposited below wave base. This will be laminated and contain a marine fauna. Between this and the barrier is a transitional zone of interlaminated sand, silt, and clay with ripples and burrows (Allen, 1967, Fig. 1).Thus as a barrier beach progrades seawards it deposits an upward-coarsening grain-size profile with a characteristic suite of sedimentary structures (Fig. 6.1).

In summary, therefore, a Recent linear shoreline consists of two high-energy zones and two low-energy zones alternating with one another parallel to the coast. In some instances the barrier sand is forced landwards against the alluvial plain. A lagoonal tidal flat complex is then absent. This situation generally occurs on stormy coasts with low input of sediment from the land (Hoyt, 1968).

The actual sedimentary sequence which is deposited from a linear clastic shoreline is a function both of sediment availability and of the rate of rise or fall of the land and sea.

All four facies described above will only be preserved where there is an abundant supply of land-derived sediment. When this happens and the shoreline is stationary all four environments deposit sequences of the four facies side by side. Such static shorelines as these are rare in the geological column because they need a very delicate balance between sedimentation and rising sea level. The Frio sand (Oligocene) of the northwest Gulf of Mexico is an example of an ancient static clastic shoreline (Boyd and Dyer, 1966). Where a high influx of sediment is accompanied by a regression of the shoreline (for which the former may be responsible) all four facies build out one above the other seaward. Such regressive sequences are similar to deltas; essentially coarsening upwards from marine shales at the base into continental sands at the top. High sediment influx accompanied by a relative rise in sea level results in a fining-upwards transgressive sequence which is a mirror image of the former type. Regressive and transgressive shorelines with all four facies preserved occur in the Tertiary Gulf Coast Province (e.g. Rainwater, 1964, Figs 7 and 8) and in Upper Cretaceous rocks of the American Rocky Mountains. The latter example is described and discussed in this chapter.

A regression of the sea from a low-lying land surface with no detritus may leave no mark in the geological record other than an unconformity. The actual shoreline may consist of a beach, or barrier beach and lagoon complex. As the sea retreats it will leave the old beach ridges stranded inland. These will be eroded and the sediment carried back to the sea to be re-deposited on the beach face. Net sedimentation above the unconformity is thus generally zero.

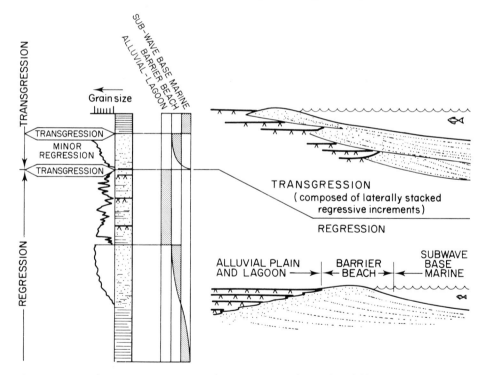

*Figure 6.2* Sedimentary section and cartoons to show the differences between regressive and transgressive shoreline sequences. Note that the transgressive sands, though they may have a sheet geometry, are made up of a series of laterally stacked regressive increments. Each individual unit is an upward-coarsening prograding sand, though of limited duration and extent.

Transgressions are more complicated than regressions. Where the transgression is rapid, and sediment in short supply, deep marine shales may unconformably overly the old land surface. When the advance of the sea is slower, and where there is sufficient input of sediment, then the four facies belts may be preserved in a mirror image of the regressive sequence. Detailed

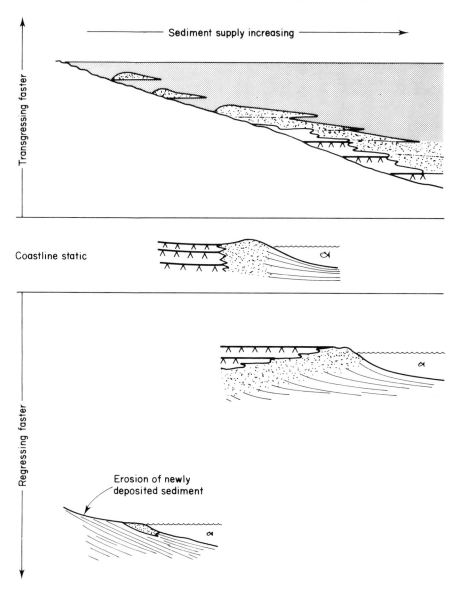

*Figure 6.3* Cartoons to show how the geometry of linear shoreline facies are controlled by the rate of sediment input and the rate of rise or fall of sea level.

examination of a transgressive section generally shows that the reflection of the 'mirror' is imperfect in so far as the marine shoreface sands are concerned. It is unusual for the barrier bar sands to grade up into the deeper water marine clays. It is more normal for them to show an upward-coarsening regressive sequence with a sharp contact with overlying marine shale. When this occurs it shows that the transgression is actually made up of a series of laterally stacked regressive increments. The sea apparently drowned a barrier beach and established a new shoreface further inland. The coast prograded seaward for a while depositing either an upward coarsening sequence or, on higher energy coasts, a sand which abruptly overlay a scoured pebble-covered sea floor. Renewed transgression drowned the second shoreface, and established a third one still further up the basin margin (Fig. 6.2).

Where sediment supply is sufficient a blanket marine sand may thus be deposited. But detailed examination will show that this is a composite sand body made up of a series of laterally stacked sands with eroded tops and upward-coarsening or sharp bases. When sediment supply is insufficient then a series of isolated shoestring sands may remain. Figure 6.3 attempts to illustrate the different types of sequence deposited in response to varying rates of sediment supply and of sea level change. These processes are discussed in many of the papers previously referred to, but see especially McCubbin (1982).

The Cretaceous sediments of the Rocky Mountain basins provide excellent examples of linear terrigenous shorelines.

## CRETACEOUS SHORELINES OF THE ROCKY MOUNTAINS, USA: DESCRIPTION

An easterly thinning clastic prism was deposited from a Cretaceous seaway which stretched across the midwest of North America from the Arctic to the Gulf of Mexico. Easterly thinning of this sequence is accompanied by a gradual decrease in grain size; coarse conglomerates derived from the rising Rocky Mountains passing eastwards through sandstones to shales. Within these rocks it is possible to recognize three major sedimentary facies which interfinger with one another laterally and are interbedded vertically (Fig. 6.4). These may be listed from west to east as follows:

(*i*) Coal-bearing facies.
(*ii*) Sheet sand facies.
(*iii*) Laminated shale facies.

These will now be described in turn.

### Coal-bearing facies
Lithostratigraphically this includes the Menefee Formation of Colorado and New Mexico and the Lance Formation and parts of the Almond and Judith River Formations of Wyoming and Montana.

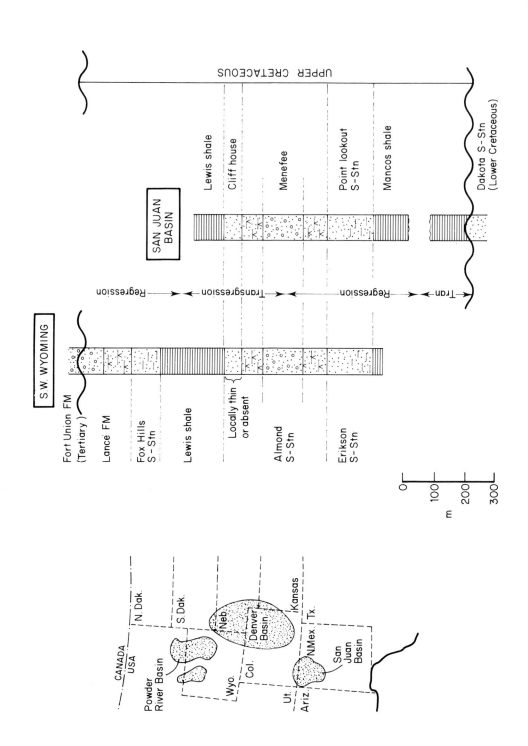

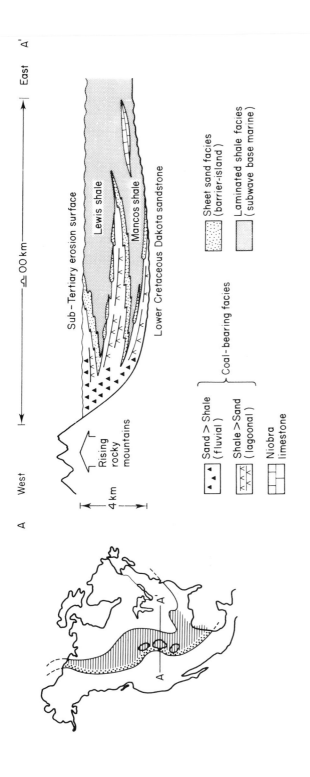

*Figure 6.4* Maps, cross-section and stratigraphic columns to illustrate the distribution of coastal deposits laid down in the Cretaceous seaway of the Rocky Mountains. Note that formation names and thickness vary from place to place. Based on sources cited at the end of this chapter.

Geometrically this facies consists of a wedge over 600 m thick in the west which splits up into a number of easterly thinning tongues when traced away from the Rocky Mountains.

In the west this facies is dominantly conglomeratic with subordinate amounts of coarse, poorly sorted sandstones and siltstones. Interbedded lavas and ashes are present, and the conglomerates and sands are often composed of volcanic detritus. These sediments show prominent channelling and cross-bedding.

Traced eastwards grain size diminishes to medium and fine cross-bedded sands with channelled bases interbedded with shales. The shales are laminated and rippled. They contain rare reptile bones and shells of the freshwater lamellibranch *Unio* and the brackish lamellibranchs *Corbula* and *Ostrea*. The oysters form reefs locally. The shales are generally dark in colour and often carbonaceous with fossil plant remains. Locally the shales and sands are interbedded with coals. These are sometimes up to 3 m thick and are commercially significant.

## Sheet sand facies

Lithostratigraphically this includes the Fox Hills Formation and the upper part of the Almond and the lower part of the Judith River Formations of Wyoming and Montana. In Colorado and New Mexico sands of this facies are represented by the Cliff House and Point Lookout Formations.

Geometrically this facies occurs in sheet-shaped units separating the coal-bearing shales of the previous facies to the west from the laminated shales of

*Figure 6.5* Regressive Point Lookout barrier sandsheet passing down transitionally into open marine Mancos shale. From Visher, 1965, Fig. 5. Reproduced from the *Bulletin of the American Association of Petroleum Geologists*, by courtesy of the American Association of Petroleum Geologists.

the third facies to the east (Fig. 6.4). Individual sand sheets are about 30 m thick and can be traced for considerable distances both along the palaeoslope (west–east) and the palaeostrike (north-south). In Colorado and New Mexico the sheets contain local thickened benches which are laterally persistent along the palaeostrike for tens of miles. Isolated sand lenticles occur interbedded with laminated shales east of the main development of sheet sands. These have shoestring geometries with north-south trends. Examples include the Eagle sandstone of Montana and the Two Wells sand lentil of New Mexico.

The petrography of the sheet-sand facies is regionally variable, ranging from glauconitic protoquartzite to carbonaceous feldspathic sand and sub-greywacke. The Gallup sand sheet in Arizona, Colorado, and New Mexico contains local heavy mineral placer deposits rich in ilmenite. Sorting is poor to moderate with considerable quantities of interstitial clay. Texture, fauna, and sedimentary structures show a regular vertical arrangement within any one sand sheet (Fig. 6.5). In the case of a regressive sand this is as follows. The upper contact of the sand with shales of the coal facies is abrupt. In at least one instance the top of the sand is channelled and infilled with an oyster-bearing siltstone (Weimer, 1961, p. 88). The upper part of the sand sequence shows the coarsest grain size and best sorting of the whole sheet. These fine sands are rarely cross-bedded and typically show low angle (5–15°) laterally persistent stratification with, generally, easterly dips (westerly in the case of the Eagle Sand shoestring). This facies often contains burrows of *Ophiomorpha* which are comparable to those produced today by the crustacean *Callianassa* on the tidal and sub-tidal parts of beaches.

These fine flat-bedded sands grade down into very fine silty sands. These are laminated, colour mottled, and sometimes burrowed. They contain the ammonites *Baculites* and *Discoscaphites* together with the lamellibranchs *Inoceramus* and *Pholadomya*. This second unit of the sheet-sand facies grades down into laminated siltstone.

Using detailed wire-line log correlations it has been shown that these sands are arranged in basinward prograding wedges. Key marker horizons, such as volcanic ash bands, demonstrate depositional topographies of up to 700 m.

## Laminated shale facies

The third facies of the Cretaceous mid-west sediments includes the Lewis, Mancos, Bearpaw, and Pierre Shale Formations. These are best developed to the east where they locally become calcareous and grade into limestones (e.g. the Niobrara Formation). Traced westwards this facies thins and splits up into a number of tongues which interfinger with the sheet-sand facies, forming the toesets of the progradational wedges of the sheet sands. Contacts are gradational with a siltstone sequence separating the shales from the very fine silty sands. In some areas the sheet sands are locally absent, and the laminated shales directly overlie the coal-bearing facies.

Lithologically this facies consists of grey laminated claystones with a fauna of ammonites and lamellibranchs similar to that recorded from the lower part of the sheet-sand facies, together with shark teeth.

## CRETACEOUS SHORELINES OF THE ROCKY MOUNTAINS, USA: INTERPRETATION

The conglomerates and coarse cross-bedded channelled sands of the extreme west were clearly deposited in a high-energy environment by fast traction currents. The finer sediments with which they are interbedded eastwards indicate lower energy conditions. The lamellibranchs in the shales suggest deposition in waters whose salinity ranged from fresh to brackish. The thin coal beds indicate intermittent swamp conditions. Considered over all, therefore, the coal-bearing facies seems to have been deposited on a piedmont alluvial plain which passed eastwards down slope into a region of swamps and brackish lagoons.

The sheet-sand facies suggests that the lagoons were restricted to the east by a higher energy environment. The ammonites and lamellibranchs of the sands indicate that they were laid down in or close to a marine environment. The vertical sequence of increasing grain size and increased sorting is typical of Recent barrier beaches (see p. 147), as also is the sequence of sedimentary structures, with the dominance of low-angle stratification in the upper part of each sand sheet. The evidence points, therefore, to a barrier beach environment for this facies. The channels at the top of each sand sequence were probably cut by tidal currents flowing between the open sea to the east and the brackish lagoons to the west.

One may therefore think of the Rocky Mountain Cretaceous shoreline as a coalesced complex of wave-dominated Nile-type delta. Alternatively one may consider the regionally extensive cyclicity of the formations as suggesting that it may be more appropriate to think of the shoreline as deposited by alternating deltaic regressive phases and barrier island transgressive phases.

The laminated shales of the eastern part of the region were clearly laid down in low-energy conditions. Their fauna indicates a marine environment. It seems most probable, therefore, that the laminated shale facies originated below wave base on a marine shelf to the east of the barrier sands.

From the preceding observations and interpretations it can be seen that these Cretaceous sediments provide a good example of a linear clastic shoreline. All four major environments found in Recent linear clastic coasts are represented in both regressive and transgressive phases of the shoreline. The coal-bearing facies represents the alluvial and lagoonal environments, the sheet sands indicate sand barriers, and the marine shale provides evidence of the open sea environment. It is clear from the repeated vertical interbedding and lateral interfingering of all the facies that deposition took place synchronously in all four environments. Several lines of evidence point to fluctuations in the rate of advance and retreat of the shoreline. The local superposition of marine shales directly on the coal-bearing facies shows that sometimes the sea transgressed too fast for barrier sands to form. This may have been caused either by shortages of land-derived sediment, or by extremely rapid rises of sea level or subsidence of the land.

The isolated sand shoestrings within the marine shales also suggest that sometimes the sea advanced so fast that barriers were no sooner formed than they were submerged to form offshore bars and shoals. The location of these was sometimes controlled by palaeohighs on the sea floor. Thus the Eagle Sand shoestring trends along the crest of an anticline. The westerly orientation of the inclined bedding of this body suggests that the shoal migrated landwards.

By contrast, the thick clean sand benches in the sheet-sand facies suggest that from time to time the shoreline was static. High barriers were then thrown up by the sea on which the sand was continually reworked and from which the clay was winnowed.

In conclusion then it can be seen that these Cretaceous sediments provide a good example of a linear clastic shoreline in which rapid deposition allowed the sediments of all four environments to be deposited during both regressions and transgressions of the coast.

## GENERAL DISCUSSION OF LINEAR TERRIGENOUS SHORELINES

As can be seen from the previous case history, clastic shorelines can be easy to identify. This is true not only for those in which all four environments are preserved during regression. It is also easy to identify the passing of a marine transgression where an unconformity is overlain by a marine-sand facies with, or without, an intervening continental sequence (Figs 6.2 and 6.3).

There is however, one particular case where confusion of diagnosis can arise. Since this can often be of economic significance it will now be discussed in some detail.

As already mentioned, prograding linear shorelines and deltas both deposit sequences which are broadly comparable. At the base is a marine shale which coarsens upwards into laminated rippled and burrowed silts. These in turn are overlain by coarser sand grade sediments, often with continental fossils and associated with coals and non-marine shales.

Both deltaic and barrier sands often make good hydrocarbon reservoirs. Deltaic distributary sands occur in down slope trending channels. Barrier sands, in contrast, are shoestrings which trend along the palaeostrike. The distinction of the precise environment of an oil-bearing shoreline sand is vital if the geometry of the reservoir is to be predicted accurately.

The confusion is particularly real in areas where barrier sands develop around active deltas, as in the Recent mouth of the Nile. Similarly, abandoned lobes of the Mississippi delta are presently being re-worked by the sea which advances over old delta topsets. Sand barriers and shoals are thrown up and as they migrate landward leave behind a veneer of marine sand on the old delta land surfaces. An ancient analogue of this process can be seen in the Mullaghmore Carboniferous delta complex in the Sligo basin of Ireland. The channel sands at the top of the upward-coarsening deltaic sequence are overlain by a clean calcareous sandstone unit with fragmented marine fossils.

These Recent and ancient examples of associated marine beach sands and delta distributary sands show how closely associated they may be, with the wave-dominated deltas like the Nile and the Niger occupying a point midway between

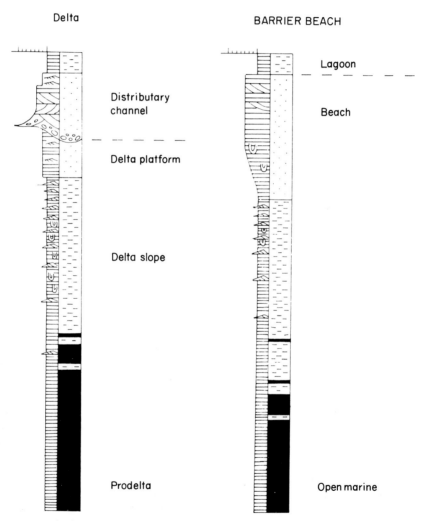

*Figure 6.6* Idealized sedimentary sections produced by prograding deltas and barrier sands. Note the close similarity between the lower parts of both sections and the fact that they both coarsen upwards. The main distinguishing features occur in the top sandstone units.

the two types. Fortunately, however, they can generally be distinguished with a few criteria. The lower marine sections are essentially identical. So also is the transitional zone of interlaminated rippled and burrowed silt and very fine sand. It is unlikely, though, that turbidites would be found in a linear shoreline sequence. Relatively slower deposition allows more time for marine currents to re-work the sediment, and gentle sea floor gradients are the rule. Turbidites, however, are not ubiquitous in deltaic shorelines (see Chapter 5). It is in the upper part of the deltaic and barrier shoreline that differences are present. A deltaic channel sand will generally have an erosional base, often overlain by a thin conglomerate. The succeeding sand will generally be cross-bedded. An upward-fining of grain size may continue into a rippled, very fine sand. Fossils in the channel will contain continental animals and plant debris, often fragmented and showing signs of transportation.

A beach sand on the other hand will have not an erosional but a transitional base passing gradually down into the laminated silty sediment beneath. The sand itself will be typically flat-bedded, with only rare isolated sets of cross-bedding. The fauna will be fragmented, but marine.

The distinguishing features of deltaic and linear shoreline sand sequences are summarized in Fig. 6.6. Davies *et al.*, (1971) have also reviewed the diagnostic criteria of barrier environments.

It can be seen, therefore, that there are certain criteria which can be used to distinguish bar sands from delta distributary sands. Only one well need be drilled through a sand to determine its origin if it is cored or if the transitional or abrupt base can be picked out on electric logs. Furthermore, if something is known of the palaeogeography of the area it may be possible to predict the orientation of the sand body.

Sometimes, however, nature may make the task of prediction easier. It has already been seen in the case of the Rocky Mountain shoreline that migrating barriers can deposit sheets of sand. Similarly, the top of a delta may consist of a sheet formed of coalesced channels. This can be seen in the deltaic sands of the Tabuk basin of Arabia (p. 128) and in the abandoned La Fourche sub-delta of the Mississippi (Kolb and Van Lopik, 1966, p. 37).

Deltas and linear clastic shorelines are comparable not only because they can both generate upward-coarsening marine shale to continental sand sequences, but also because these are often cyclically repeated. This is shown by the Rocky Mountain Cretaceous sediments. Cyclic linear clastic shoreline sequences are particularly characteristic of the Tertiary Era. They have been described from rocks of this age from as far apart as the Anglo-Paris Basin (Stamp, 1921), the Gulf Coast of Louisiana and Texas (Bornhauser, 1947; Lowman, 1949; Fisher, 1964; MacNeil, 1966), and Sumatra (Dufour, 1951).

Unlike deltas, barrier shorelines contain no sedimentary process which can act as a built-in cycle generator. It is generally agreed, therefore, that such cycles are due to eustatic and/or tectonic causes (Duff, Hallam, and Walton, 1967, p. 187; Tanner, 1968).

# ECONOMIC SIGNIFICANCE OF LINEAR TERRIGENOUS SHORELINES

Since linear shorelines deposit clean porous sands around marine basins they are often prolific oil and gas producers. The barrier sands obviously provide the best potential reservoirs with both the marine and the lagoonal shales being potential hydrocarbon source rocks. It is, therefore, very important to be able to recognize bar and beach sands and to make predictions of their geometry and trend. The criteria for distinguishing such sands from deltaic channels has been discussed in the previous section. The significance of this distinction is self-explanatory.

Though at any one time a barrier or beach is a linear environment running parallel to the shore, the geometry of the sand that is actually deposited can be quite varied. There are three main kinds:

(1) Regressive sand sheets,
(2) transgressive sand sheets,
(3) shoestring barrier sands.

These will now be described in turn.

## Regressive sand sheets

The Cretaceous shoreline sediments of the Rocky Mountain region contain good examples of regressive (as well as transgressive) sand sheets. These have established gas reserves in excess of 20 trillion cubic feet with some oil as well. Though some production comes from the continental facies, the bulk is from the marine barrier sands. These sands have sheet geometries, and are, therefore, easy to locate. It is not so easy though to find parts with good reservoir properties. Due to the clay matrix, porosity and permeability values are low. Optimum reservoir characteristics are found where there is secondary fracture porosity and where the sands are particularly well-sorted and clay-free. The second of these features is found in two situations. It occurs at the top of each sand sequence, where it is easy to find, and in the thick sand benches. Since the latter occur in narrow belts often only two or three miles wide, they are not always easy to locate, nor, once found, is their regional trend simple to predict. The Cretaceous shoreline did not extend in a straight line north to south from Canada to the Gulf of Mexico. Like Recent coasts, it had bays, capes, and spits, so that locally the thick barrier sand benches have trends varying from northwest to northeast. Fortunately these strata are gently folded, and due to erosion the barrier sands crop out intermittently at the surface. In such situations the thick porous sands can be located and their trend predicted underground.

Evans (1970) has shown how the Cretaceous Viking bar sands of Saskatchewan were deposited as a series of linear imbricately arranged bodies. The sands strike approximately parallel to the basin margin and the direction of imbrication is towards the basin centre.

## Transgressive sand sheets

The geometry of transgressive sands is largely controlled by the shape of the unconformity over which the sea advances. Where this is a planar surface, the overlying sand will have a sheet geometry. Though the sand is a sheet, oil reservoirs within it will be linear in plan, trending parallel to the shoreline. Their up dip limit will be controlled by the sand pinch out and their down dip extension will be bounded by the oil: water contact (Fig. 6.7a). Oil reservoirs occur in basal transgressive sands of both marine and non-marine origin (e.g. the Lower Cretaceous Cutbank sand of Montana and the Pliocene Quirequire field of Venezuela respectively (Leverson, 1967, pp. 336–7)).

( a )                                        ( b )

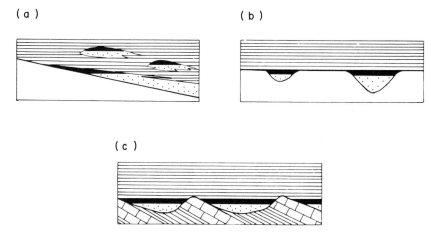

( c )

*Figure 6.7* Diagrams of some types of stratigraphic oil and gas reservoirs in terrigenous linear shoreline deposits. Oil; black; basement; blank; sand; stippled; shale cap rock; ruled. a. Palaeodip section showing reservoirs in basal transgressive sand sheet and isolated barrier sand shoestrings. b. Palaeostrike section showing oil accumulation in channel sands. c. Reservoirs in sands infilling strike valleys cut in softer beds of alternating hard and soft strata.

More complicated reservoir geometries occur where clastic shorelines transgress over irregular land surfaces. Basal sands, both marine and non-marine, are then laid down in the low-lying valleys, but are often absent over the hills. The study of buried land surfaces, termed palaeogeomorphology, has been discussed from an economic angle by Martin (1966). Two main kinds of stratigraphic reservoirs may be found in sands overlying irregular topographies. Old river channels may trend seawards down the palaeoslope. As the shoreline advances the fluviatile sands are buried under marine shales. Oil and gas reservoirs may then form in linear pools parallel to the palaeoslope (Fig. 6.7b). There are many instances of such reservoirs. Good examples occur in basal Cretaceous sands of Canada (Martin, 1966). A special case of this type of reservoir is where oil accumulates in river terrace sands buried under shale (Conybeare, 1964).

The second type of hydrocarbon accumulation on irregular transgressive surfaces is where sands are deposited in strike valleys perpendicular to the consequent palaeodrainage surface. These form where the rocks beneath the unconformity consisted of interbedded, gently dipping hard (scarp-forming), and soft (valley-forming) strata (Fig. 6.7c). Oil reservoirs in strike valley sands occur in the basal Cretaceous of Alberta and Saskatchewan (Martin, 1966) and in basal Pennsylvanian Cherokee sands of Kansas and Oklahoma (Busch, 1961).

Obviously a considerable understanding not only of sedimentology but also of geomorphology is needed to predict the location of hydrocarbon reservoirs in the basal sands of transgressive clastic shorelines.

## Shoestring barrier sands

Shoestring sands of clastic shorelines are of two main types. One occurs as barriers separating marine from non-marine shale. The second type are shoestrings entirely enclosed in marine shale. The origin of such sands has been discussed in the light of data from Recent shelf sea sand bars by Off (1963). Prolific oil production comes from Tertiary barrier shoestring sands of the Gulf Coast province of Texas and Louisiana. Though many of these barriers form transgressive and regressive sheets (Rainwater, 1964, Figs 7 and 8), shoestrings also occur separating marine from non-marine shale. The Frio sand (Oligocene) is a good example, being up to 1500 m thick and 40 km wide. Oil production extends for several hundred kilometres along the palaestrike (Burke, 1958; Boyd and Dyer, 1966). Similar examples occur in the Middle Vicksburg (Lower Oligocene) of the same region. Here barrier sand shoestrings lie to the northeast and southwest of a deltaic complex separating lagoonal shales to the northwest from open marine shales to the southeast. These shoestrings are about 20 m thick and 5 m wide (Gregory, 1964).

There are many well-documented cases of oil and gas production from shoestring sands entirely enclosed within marine shale. Particularly good examples are found in Pennsylvanian sediments of Oklahoma and Kansas. Individual bars are three or more kilometres in width and some 20 m thick. In plan the shoestrings are often arranged *en echelon*, though trends, both linear and curved, can be traced from upwards of 75 kilometres (Busch, 1961; Leverson, 1967, Figs 7–14). Within the Enid empayment of the Anadarko basin of Oklahoma oil accumulations in concentric marine shoestrings are associated with reservoirs in basinward trending channel sands (Withrow, 1968).

Similar fields in offshore bars occur in the Silurian Clinton Sands and Late Devonian: Early Mississippian of Pennsylvania and West Virginia (Potter and Pettijohn, 1963, Figs 9.3 and 9.4).

Linear clastic shorelines are not only of economic importance because of their ability to trap oil and gas. They may contain coals formed in the swamps and lagoons behind barriers, as in the Cretaceous Mesaverde Formation of the Rocky Mountains. Carboniferous coal measures, though largely deltaic, also contain coals formed behind barrier sands (p. 130). Beach sands can contain winnowed valuable detrital minerals, such as the Recent ilmenite-rich beach sands of India, Ceylon, and New Zealand (Deer, Howie, and Zussman, 1962, p. 31). Ilmenite-rich Cretaceous beach deposits of the Gallup Sandstone in New Mexico are an ancient analogue (Murphy, 1956).

The copper belt of Zambia is another example of a linear clastic shoreline of considerable economic significance. Here a Pre-Cambrian sea advanced across a rugged granite land surface with fluviatile and eolian sands. Pyrite, chalcopyrite, bornite, and chalcocite occur in both the fluviatile sands and in overlying marine shales, sandstones, and dolomites. Mineralization is principally developed in bays and drowned valleys of the shoreline. Regardless of whether the minerals formed at the time of deposition of the sediment, or later, sedimentological studies have an important part to play in unravelling the palaeogeography and environment of such deposits, thus helping to locate ore bodies and to predict their geometry (Mendelsohn, 1961; Garlick, 1969; Fleischer *et al.*, 1976).

It can be seen therefore that the study of linear clastic shorelines can be very profitable.

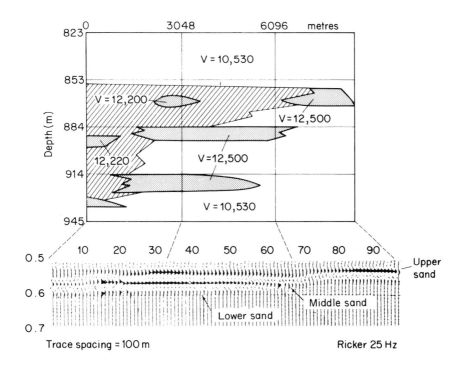

*Figure 6.8* Geological model (upper) and seismic response (lower) for Cretaceous Belly River bar sands, Alberta. (From Meckel and Nath 1977, by courtesy of the American Association of Petroleum Geologists).

## SUB-SURFACE DIAGNOSIS OF BARRIER SANDS

Because of their potential as hydrocarbon reservoirs great attention has been paid to the sub-surface diagnosis of barrier sands (Davies *et al.*, 1971).

Geometrically these deposits may have either shoestring or sheet geometries. Considered in detail the sheets are composed of a series of laterally stacked discrete beach increments, as shown by the Viking Cretaceous Sands of Canada already mentioned (Evans, 1970). Prograding barrier bar sands generate an upward-coarsening grainsize profile. They normally pass down gradationally into marine shales, and commonly have an abrupt top. The upper surface of a barrier bar sand sometimes has sufficient acoustic impedance to generate a seismic reflector. (This is of course the reverse situation to upward-fining channels where a reflector is to be anticipated at the base p. 73.) Figure 6.8 shows the seismic signature produced from modelling studies of Cretaceous Belly River bar sands from Alberta. These are a northward continuation of the Rocky Mountain Cretaceous sands of the USA discussed earlier in the chapter. It is sometimes possible to use seismic data to map individual sandbars and the

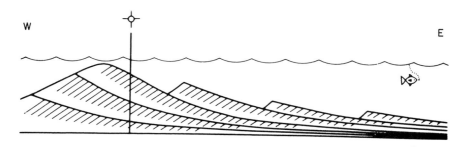

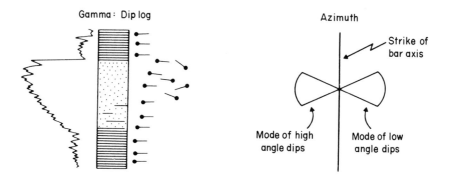

Gamma and dip motifs for barrier/bar sands

*Figure 6.9* Gamma log and dipmeter motifs for barrier bar sand bodies.

corrugated upper surface of barrier sand sheets. Thus environmental diagnosis may be made prior to drilling.

Lithologically marine sands can often be recognized in well samples by their textural and mineralogical maturity, by the presence of glauconite and shell debris, and by the absence of mica and carbonaceous detritus.

Cores through barrier sand bars may show the suite of sedimentary structures shown in Fig. 6.6 (right) associated with an upward-coarsening grain-size profile. Where cores are not available this motif may nonetheless be clearly shown by gamma or SP logs.

The typical dip pattern for regressive barrier sands is an upward-increasing blue motif. When plotted on a rose diagram a bimodal bipolar pattern may be seen. The dips of the blue motif represent low angle seaward dipping beds, while the opposed mode with higher dip is due to onshore directed foresets (Fig. 6.9). Even if this bipolar pattern is indecipherable the direction of dip of the blue curve can be used to find the strike of the bar, and the direction in which the axis lies. This is extremely useful information to know both in the quest for bar sand stratigraphic traps, and for regional palaeogeographic studies.

## REFERENCES

The account of the Cretaceous shoreline east of the Rocky Mountains was based on:

Asquith, D. O., 1970. Depositional topography and major marine environments, late Cretaceous, Wyoming. *Bull. Amer. Assoc. Petrol. Geol.*, **54**, 1184–224.

Davies, D. K., Ethridge, F. G., and Berg, R. R., 1971. Recognition of barrier environments. *Bull. Amer. Assoc. Petrol. Geol.*, **55**, 550–65.

De Graaff, F. R. Van, 1972. Fluvial-Deltaic facies of the Castlegate sandstone (Cretaceous), east-central Utah. *J. Sediment. Petrol.*, **42**, 558–71.

Evans, W. E., 1970. Imbricate linear sandstone bodies of Viking formation in Dodsland-Hoosier area of southwestern Saskatchewan, Canada. *Bull. Amer. Assoc. Petrol. Geol.*, **54**, 469–86.

Hollenshead, C. T., and Pritchard, R. L., 1969. Geometry of producing Mesaverde Sandstones, San Juan basin. In: *The Geometry of Sandstone Bodies*. (Eds J. A. Peterson and J. C. Osmond) Amer. Assoc. Petrol. Geol. pp. 98–118.

Mackenzie, D. B., 1975. Tidal flat deposits in Lower Cretaceous Dakota Group near Denver, Colorado. In: *Tidal deposits*. (Ed. R. N. Ginsburg) Springer-Verlag, New York. pp. 117–25.

McCubbin, D. G., 1982. Barrier Island and Strand Plain Facies. In: *Sandstone Depositional Environments*. (Eds P. A. Scholle and D. Spearing) Amer. Assoc. Petrol. Geol. Tulsa. 247–80.

Miller, D. N., Barlow, J. A., and Haun, J. D., 1965. Stratigraphy and petroleum potential of latest Cretaceous rocks, Bighorn basin, Wyoming. *Bull. Amer. Assoc. Petrol. Geol.*, **49**, 227–85.

Pike, W. S., 1947. Intertonguing marine and non-marine Upper Cretaceous deposits of New Mexico, Arizona, and Southwest Colorado. *Geol. Soc. Amer. Mem.* No. 24.

Shelton, J. W., 1965. Trend and genesis of lowermost sandstone unit of Eagle Sandstone at Billings, Montana. *Bull. Amer. Assoc. Petrol. Geol.*, **49**, 1385–97.

Siemers, C. T., 1976. Sedimentology of the Rocktown Channel Sandstone, upper part of the Dakota formation (Cretaceous), central Kansas. *Jour. Sedim. Petrol.*, **46**, 97–123.

Viele, G. W., and Harris, F. G., 1965. Montana Group Stratigraphy, Lewis and Clark County, Montana. *Bull. Amer. Assoc. Petrol. Geol.*, **49**, 379–417.

Weimer, R. J., 1961. Spatial dimensions of Upper Cretaceous Sandstones, Rocky Mountain Area. In: *Geometry of Sandstone Bodies.* Ed. J. A. Peterson and J. C. Osmond) Amer. Assoc. Petrol. Geol. pp. 82–97.

——, 1966. Patrick Draw Field, Wyoming. *Bull. Amer. Assoc. Petrol. Geol.*, **50**, 2150–75.

——, and Haun, J. D., 1960. Cretaceous stratigraphy, Rocky Mountain Region, USA. *Int. Geol. Congr. Norden*, pt. 12, pp. 178–84.

——, and Land, C. B., 1975. Maestrichtian deltaic and interdeltaic sedimentation in the Rocky Mountain region of the United States. In: The Cretaceous System in the Western Interior of North America. (Ed. W. G. E. Caldwell) *Geol. Assn. Canada*, Sp. Pub. 13, pp. 633–66.

Other references cited in this chapter:

Aharoni, E., 1966. Oil and gas prospects of Kurnub Group (Lower Cretaceous) in Southern Israel. *Bull. Amer. Assoc. Petrol. Geol.*, **50**, 2388–403.

Allen, J. R. L., 1967. Depth indicators of clastic sequences. In: Depth indicators in marine sedimentary environments. (Ed. A. Hallam) *Marine Geology*, Sp. Issue, **5**, No. 5/6, pp. 429–46.

Armstrong-Price, W., 1963. Patterns of flow and channelling in tidal inlets. *J. Sediment. Petrol.*, **33**, pp. 279–90.

Bender, F., 1968. *Zur geologie von Jordanien.* Beitr. Reg. Geol. Erde. Bd. 7, Berlin.

Bornhauser, M., 1947. Marine sedimentary cycles of Tertiary in Mississippi Embayment and Central coast area. *Bull. Amer. Assoc. Petrol. Geol.*, **31**, 696–712.

Boyd, D. R., and Dyer, B. F., 1966. Frio Barrier bar system of South Texas. *Bull. Amer. Assoc. Petrol. Geol.*, **50**, 170–8.

Burke, R. A., 1958. Summary of oil occurrence in Anahuac and Frio formations of Texas and Louisiana. *Bull. Amer. Assoc. Petrol. Geol.*, **42**, 2935–50.

Busch, D. A., 1961. Prospecting for stratigraphic traps. In: *Geometry of Sandstone Bodies.* (Eds J. A. Peterson and J. C. Osmond) Amer. Assoc. Petrol. Geol. pp. 220–32.

Conybeare, C. E. B., 1964. Oil accumulation in alluvial stratigraphic traps. *Australian Oil and Gas. Jl.* August 1964, p. 5.

Davies, R. A. (Ed.). 1978. *Coastal Sedimentary Environments.* Springer-Verlag, Berlin. p. 420.

Davis, R. A., and Ethington, R. L. (Eds), 1976. Beach and nearshore sedimentation. *Soc. Econ. Paleont. Miner*, Sp. Pub. No. 24 Tulsa.

Deer, W. A., Howie, R. A., and Zussman, J., 1962. *Rock-forming Minerals. Vol. 5, Non-Silicates.* Wiley, N. Y. p. 370.

De Raaf, J. F. M., and Boersma, J. R., 1971. Tidal deposits and their sedimentary structures. *Geol. Mijnb.* **50**. 479–504.

Duff, P. L., McL. D., Hallam, A., and Walton, E. K., 1967. *Cyclic Sedimentation.* Elsevier, Amsterdam. p. 280.

Dufour, J., 1951. Facies shift and isochronous correlation. *Proc. 3rd World Petrol. Cong.* Netherlands. Section 1, pp. 428–37.

Evans, G., 1965. Intertidal flat sediments and their environments of deposition in the Wash. *Quart Jl. Geol. Soc. Lond.*, **121**, 209–45.

Fisher, W. L., 1964. Sedimentary patterns in Eocene cyclic deposits, northern Gulf Coast region. *Kansas Geol. Surv. Bull.* 169, pp. 151–70.

Fleisher, V. D., Garlick, W. G. and Haldane, R. 1976. Geology of the Zambian Copperbelt. In: *Handbook of Strat bound and Stratiform Ore Deposits.* Vol. 6. (Ed. K. H. Wolf Elsevier, Amsterdam. pp. 223–52.)

Garlick, W. G., 1969. Special features and sedimentary facies of stratiform sulphide deposits in arenites. In: *Sedimentary Ores Ancient and Modern (Revised).* (Ed. C. H. James) Sp. Pub. No. 1, Geol. Dept. Leicester Univ., U.K. pp. 107–69.

Ginsberg, R. N. (Ed.), 1975. *Tidal deposits.* Springer-Verlag, New York. p. 428.

Gregory, J. L., 1966. A Lower Oligocene delta in the subsurface of south-eastern Texas. In: *Deltas.* (Eds M. L. Shirely and J. A. Ragsdale) Houston Geol. Soc., pp. 231–28.

Guilcher, A., 1970. Symposium on the evolution of shorelines and continental shelves in their mutual relations during the Quaternary. *Quaternaria*, **12**, p. 229.

Hails, J., and Carr, A. P. (Eds), 1975. *Nearshore sediment dynamics and sedimentation.* J. Wiley & Sons, London, p. 316.

Hammuda, O. S., 1969. Jurassic and Lower Cretaceous Rocks of Central Jabal Nefusa, Northwestern Libya. *Petrol. Explor. Soc. Libya Guidebook.* p. 74.

Hobday, D. K., and Erikson, K. A. (Eds), 1977. Tidal sedimentation. *Sed. Geol.*, **18**, (Sp. Issue) pp. 1–287.

Hoyt, J. H., 1968. Genesis of sedimentary deposits along coasts of submergence. *Rept. 23, Int. Geol. Cong. Prague*, p. 231.

——, and Henry, V. J., 1967. Influence of island migration on barrier island sedimentation. *Bull. Geol. Soc. Amer.*, **78**, 77–86.

——, and Weimer, R. J., 1963. Comparison of modern and ancient beaches, Central Coast. *Bull. Amer. Assoc. Petrol. Geol.*, **47**, 529–31.

——, ——, and Henry, V. J., 1964. Late Pleistocene and Recent sedimentation, central Georgia Coast, USA. In: *Deltaic and Shallow Marine Deposits.* (Ed. L. M. J. U. Van Straaten) Elsevier, Amsterdam. pp. 170–6.

King, C. A. M., 1972. *Beaches and coasts* (2nd Edn.) Arnold, London. p. 570.

Klein, G. de V., 1967. Paleocurrent analysis in relation to modern marine sediment dispersal patterns. *Bull. Amer. Assoc. Petrol. Geol.*, **51**, 366–82.

Kolb, C. R., and Van Lopik, J. R., 1966. Depositional environments of the Mississippi River deltaic plain, southeastern Louisiana. In: *Deltas.* (Eds M. L. Shirley and J. A. Ragsdale) Houston Geol. Soc., pp. 17–62.

Leverson, A. I., 1967. *Geology of Petroleum.* Freeman & Co., San Francisco. p. 724.

Lowman, S. A., 1949. Sedimentary facies in Gulf Coast. *Bull. Amer. Assoc. Petrol. Geol.*, **33**, 1039–97.

MacNeil, F. S., 1966. Middle Tertiary Sedimentary Regimen of Gulf Coast Region. *Bull. Amer. Assoc. Petrol. Geol.*, **50**, 2344–65.

Martin, R., 1966. Paleogeomorphology. *Bull. Amer. Assoc. Petrol. Geol.*, pp. 2277–311.

Meckel, L. D. and Nath, A. K., 1978. Geological Considerations for Stratigraphic Modelling and Interpretation. In: *Seismic Stratigraphy-applications to hydrocarbon Exploration.* (Ed. C. E. Payton). Amer. Assoc. Petrol. Geol. Mem. 26. 417–38.

Mendelsohn, F., 1961. *The Geology of the Northern Rhodesian Copperbelt.* Macdonald, London.

Meyer, R. E. (Ed.), 1972. *Waves on beaches and resulting sediment transport.* Academic Press, London.

Murphy, J. F., 1956. Preliminary report on Titanium-bearing Sandstone in the San Juan basin and Adjacent Areas in Arizona, Colorado, and New Mexico. *U.S. Geol. Surv. Open File Report*, No. 385.

Narayan, J., 1971. Sedimentary structures in the Lower Greensand of the Weald, England Bas-Boulounnais, France. *Sedimentary Geology*, **6**, 73–109.

Off, T., 1963. Rhythmic linear sand bodies caused by tidal currents. *Bull. Amer. Assoc. Petrol. Geol.*, **47**, 324–41.

Oomkens, E., 1967. Depositional sequences and sand distribution in a deltaic complex. *Geol. Mijnb.*, **46**, 265–78.

Pannekoek, A. J., 1956. *Geological History of the Netherlands.* Govt. Printing Office, the Hague. p. 147.

Potter, P. E., and Pettijohn, F. J., 1963. *Paleocurrents and Basin Analysis.* Springer-Verlag, Berlin. p. 296.

Rainwater, E. H., 1966. The Geologic Importance of deltas. In: *Deltas.* (Eds M. L. Shirley and J. A. Ragsdale) Houston Geol. Soc., pp. 1–16.

Selley, R. C., 1968. Nearshore marine and continental sediments of the Sirte Basin, Libya. *Quart. Jl. Geol. Soc. Lond.*, **124**, 419–60.

Selley, R. C., 1982. *Introduction to Sedimentology.* 2nd edn. Academic Press Inc, London. p. 417.

Shelton, J. W., 1965. Trend and genesis of lowermost sandstone unit of Eagle Sandstone at Billings, Montana. *Bull. Amer. Assoc. Petrol. Geol.*, **49**, 1385–97.

Stamp, L. D., 1921. On cycles of sedimentation in the Eocene strata of the Anglo-France-Belgian Basin. *Geol. Mag.*, **58**, 108–14, 146–57, and 194–200.

Steers, J. A., 1971. *Introduction to coastline development.* Macmillan, London. p. 229.

Sweet, K., Klein, G. de V., and Smit, D. E., 1971. A Cambrian tidal sand body — the Eriboll Sandstone of northwest Scotland: an ancient:recent analog. *J. Geol.*, **79**, 400–15.

Swift, D. J. R. and Palmer, H. D. (Eds). 1978. *Coastal Sedimentation.* Dowden, Hutchinson and Ross, Stroudsburg. p. 339.

Tanner, W. F. (Ed.), 1968. Tertiary Sea-level Fluctuations. *Palaeogeography, Palaeoclimatology, Palaeoecology*, Sp. Issue. p. 178.

Weimer, R. J., and Hoyt, J. H., 1964. Burrows of *Callianassa major* SAY, geologic indicators of littoral and shallow neritic environments. *Jl. Pal.*, **38**, 761–7.

Withrow, P. C., 1968. Depositional environments of Pennsylvanian Red fork sandstone in northeastern Anadarko Basin, Oklahoma. *Bull. Amer. Assoc. Petrol. Geol.*, **52**, 1638–54.

Van Straaten, L. M. J. U., 1959. Minor structures of some recent littoral and neritic sediments. *Geol. Mijnb.*, **21**, 197–216.

# 7 Mixed Terrigenous: carbonate shorelines

## INTRODUCTION

Mixed carbonate:terrigenous shorelines are defined as those in which carbonate deposition occurs so close to the land that it contributes, not just to the open sea facies, but also to the shoreline deposits themselves (see p. 185).

These conditions can be brought about by three factors acting singly or in concert. Low input of terrigenous sediment to the shoreline may be due to low runoff or, if the hinterland is low-lying, low sediment availability. Thirdly, if the shoreline itself has a very gentle seaward gradient it will have an extremely broad tidal zone and an extremely wide development of the facies belts paralleling the shore. In such instances, terrigenous sediment may be dumped around river mouths in estuarine tidal flats. There may be insufficient current action to re-work these deposits and carry sand out to the barrier zone. In this case, onshore currents may pile up bars of carbonate sand far to seaward of the river mouths.

A particularly good example of a mixed carbonate:terrigenous shoreline occurs in Miocene rocks of the Sirte basin, Libya. All three of the factors outlined above seem to have operated in this instance. This case history will now be described.

## A LIBYAN MIOCENE SHORELINE: DESCRIPTION AND DISCUSSION

The Sirte basin developed in Upper Cretaceous time as a southerly extension from the Tethys geosyncline by block-faulting and subsidence of a part of the Sahara Shield.

By the Miocene Period the embayment had largely been infilled by carbonates and shales. Through late Tertiary time a gradual regression of the sea was interrupted by minor marine transgressions. One of these in Lower Miocene time (Aquitanian-Burdigalian) deposited the shoreline sediments now to be described. These have been studied in the region of Marada Oasis and the Jebel Zelten (Figs 7.1 and 7.2).

   In the western part of the region, an unconformity between Miocene and
Oligocene sediments crops out at the surface. This seems to have been subjected
to intensive sub-aerial weathering. The Oligocene limestones immediately
beneath the unconformity are extensively oxidized, replaced by gypsum, and
penetrated by the roots of trees which grew on the pre-Miocene land surface.
The succeeding Miocene shoreline deposits, some 200 m thick, are overlain by
a later Miocene formation. The contact with this is unconformable in the Jebel
Zelten, but becomes transitional when traced northwards towards the centre
of the basin. This contact has a present-day northerly dip of about 1:1000. Used
as a datum, this surface shows that the underlying Miocene sediments were
previously folded into a gentle east-west trending anticline with a culmination
over the Zelten oil field.

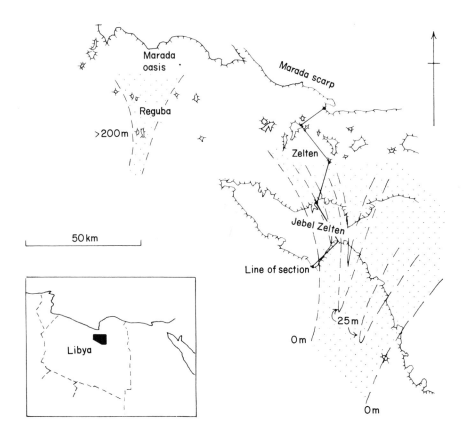

*Figure 7.1*  Location map of the Miocene shoreline of Marada Oasis and the Jebel
Zelten, Libya. Approximate areal extent of major calcareous sand complexes shown
stippled. These are aligned along two Miocene synclinal axes which previously, and
subsequently, were positive trends.

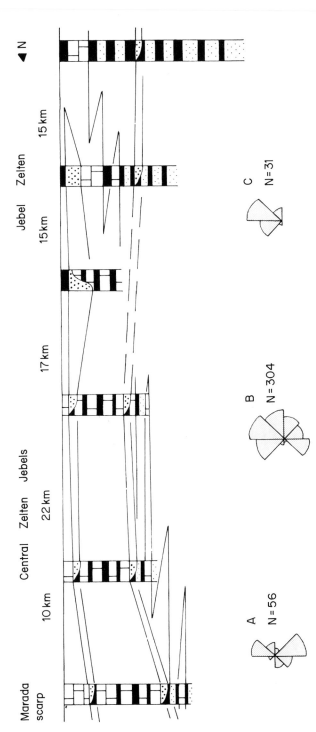

*Figure 7.2* Diagrammatic measured sections indicating facies changes across the Miocene shoreline between the Marada scarp and the Jebel Zelten. Datum is post-Marada erosion surface. Right hand section 180 m thick. Key: open marine and barrier limestones: blocks; lagoonal and intertidal shales and sands: black; fluviatile sands: dense stipple; estuarine calcareous sandstones: sparse stipple; fluviatile sands and sands: black; fluviatile sands: dense stipple. Grain size and frequency of cross-bedding in limestones increases over the crest of the anticline in the central Zelten Jebels. Azimuth diagrams show crossbedding dip orientations North vertical. Left: limestones show bipolar pattern possibly due to tidal currents; southerly directed major mode indicates dominance of onshore transport. Centre: calcareous sandstones show bipolar pattern with major mode directed offshore, probably due to deposition by tidal currents in estuarine channels. Right: unimodal seawards orientation of cross-bedding shown by fluviatile sands.

The Miocene shoreline deposits are composed of a diversity of repeatedly interbedded facies which may be listed as follows:

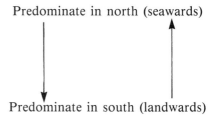

1.  Skeletal limestone facies
2.  laminated shale facies
3.  interlaminated sand and shale facies
4.  cross-bedded sand and shale facies
5.  calcareous sandstone channel facies

Predominate in north (seawards)

Predominate in south (landwards)

North-south radiating shoestring complexes

Each facies will now be described and interpreted in turn.

## Skeletal limestone facies
### Description
The skeletal limestones compose nearly the complete 200 m section exposed in the north. Traced southwards they die out due to interfingering with shales and sands. Petrographically these are poorly sorted medium- and coarse-grained packstones, composed of a framework of bioclastic debris and pellets with interstitial micrite and sparite. They still have a high intergranular porosity.

*Figure 7.3*  Wind-eroded surface showing large-scale cross-stratification in offshore bar detrital limestone facies. From Selley, 1968. Plate 21(a), by courtesy of the Geological Society of London.

Finer calcarenites, calcilutites, and calcirudites are also present. There are considerable regional variations in the grain size of this facies; coarser calcarenites being concentrated over the crest of the Zelten anticline, while finer carbonate sands and muds are present to north and south.

The coarser-grained limestones are flat-bedded or cross-bedded in isolated sets about a metre thick. In plan view it can be seen that both tabular planar and trough foresets are present, often on a vast scale. Individual planar foresets can be traced along strike for over 100 m. Troughs are up to 50 m wide (Fig. 7.3). The orientation of these structures indicates deposition from alternately onshore (south) and offshore (north) flowing currents, though with a dominance of onshore transport. The finer limestones are generally massive, and often bioturbated. At many points limestone beds are cross-cut by large channels, several metres deep. These are infilled by calcarenites similar to those in which they are cut. The channel margins and floors are often lined with re-worked, bored limestone cobbles and boulders up to a metre in diameter.

The limestones are largely composed of a diverse biota in all stages of preservation from entire and obviously *in situ*, to highly comminuted. This includes calcarous algae, bryozoa, corals, lamellibranchs (such as oysters and scallops), gastropods, echinoids, foraminifera (including miliolids and peneroplids), and a diverse suite of trace fossils (including *Ophiomorpha*).

### Interpretation

Detailed comparison of the fossils with Recent forms indicate that they grew in a variety of habitats ranging in depth from 0–50 m, from fully marine to slightly brackish salinites and from low- to high-energy bottom conditions.

The coarse bioclastic cross-bedded limestones were probably deposited in turbulent conditions. Comparison with Recent carbonate sands suggests that

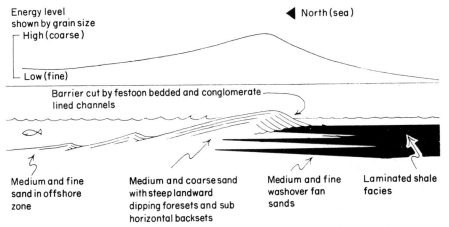

*Figure 7.4* Illustrative of the presumed origin of offshore bars in the skeletal limestone facies. From Selley, 1968, Fig. 6, by courtesy of the Geological Society of London.

they originated from migrating offshore bars and shoals. Recent examples of these deposit foresets on their steep slopes and sub-horizontal bedding on their gentle backslopes (see McKee and Sterrett, 1961, p. 62, Fig. 5). The troughs and channels cross-cutting the shoals may be analogous to those which tidal currents cut through Recent carbonate barriers (e.g. Jindrich, 1969). This interpretation is supported by their bipolar palaeocurrents (Fig. 7.2).

Steep channel margins and intraformational limestone conglomerates testify to penecontemporaneous diagenesis comparable to the formation of 'beachrock' by the sub-aerial exposure of Recent carbonate beaches (e.g. Ginsburg, 1953).

Considered over all, therefore, it seems most probable that the coarser limestones were deposited by shoreward migrating offshore bars which, intermittently exposed above sea level, became barrier islands.

The finer-grained, more generally massive and burrowed calcarenites point to deposition in less turbulent conditions largely below wave base, and perhaps in depths as great as that suggested by the fossils (Fig. 7.4).

## Laminated shale facies
### Description
Beds of shale 1 or 2 m thick occur across the whole region, interbedded with limestones in the north and sands in the south. They are grey or green in colour,

*Figure 7.5*  Oyster shell bank with valves in vertical growth position. From Selley, 1968, Plate 21(b), by courtesy of the Geological Society of London.

often highly calcareous and contain thin white calcilutite bands. Internally, the shales are laminated throughout and occasionally rippled and burrowed.

Within the shales, and separating them from limestones beneath, it is common to find oyster reefs. These are often up to a metre thick and composed of the *in situ* shells of long thin oysters, oriented vertically with their umbos pointed downwards (Fig. 7.5). Apart from the oysters, this facies contains rare bryozoa and calcareous algae and plant debris.

*Interpretation*
The lamination and fine grain size of this facies indicate deposition out of suspension in a low-energy environment. The fossils suggest a range of salinity from normal marine to brackish. These conditions could be fulfilled either in relatively deep water below wave base, or in shallow water sheltered from the open sea. The brackish element of the fauna and the way in which the shales interfinger seaward with carbonate shoals suggest that the latter alternative is the correct one. The laminated shale facies may, therefore, be attributed to a lagoonal environment.

## Interlaminated sand and shale facies
*Description*
The third facies of this Miocene shoreline occurs in lenticular units two or three metres thick, interbedded with all the other facies. It is erratically distributed through the region, being most common in the Jebel Zelten and lower parts of the section exposed to the north. It consists of sands and interlaminated sand

*Figure 7.6* Interlaminated very fine sand and shale with *Diplocraterion* burrows. From Selley, 1968. Plate 22(b), by courtesy of the Geological Society of London.

and shales, with rare thin lignites and rootlet beds. The sands are fine-grained, well-sorted, and argillaceous. Internally they are massive, laminated, or micro-crosslaminated. Associated with the sands are delicately interlaminated very fine sands, silts, and clays. These are typically rippled and highly burrowed with vertical, often U-shaped, sand-filled tubes attributable to the ichnogenus *Diplocraterion* (Fig. 7.6).

*Figure 7.7*  Fining upward tidal-creek sequences. (1) basal erosion surface, (2) channel-lag conglomerate, (3) cross-bedded coarse-medium sandstone of channel bar, (4) cross-laminated fine sand of point bar, (5) laminated siltstone channel fill. From Selley, 1968, Plate 23, by courtesy of the Geological Society of London.

Cross-cutting these sediments are largely curvaceous channels up to 20 m deep and 80 m wide. The channels are infilled by sediments showing a regular sequence of sedimentary structures and a vertical decrease of grain size (Fig. 7.7). The scoured channel floor is overlain by a conglomerate which grades up into several metres of crossbedded calcareous sand. This is overlain by interlaminated rippled and burrowed very fine sand, silt, and clay beds which dip obliquely off the channel walls and are succeeded transitionally by horizontally laminated shale (Fig. 7.8).

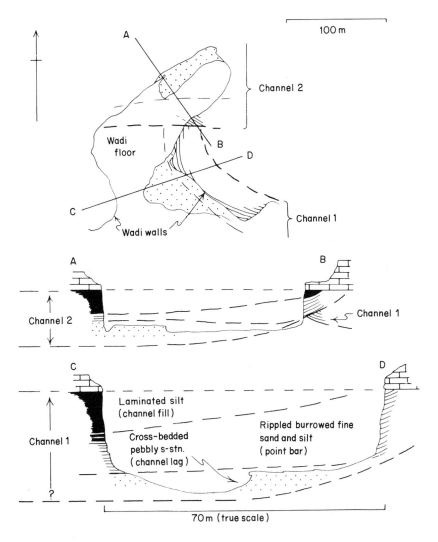

*Figure 7.8* Tidal creek channels. From Selley, 1968, Fig. 10, by courtesy of the Geological Society of London.

*Interpretation*

The over-all fine grain size of this facies indicates deposition in a relatively low energy environment. Ripples of sand and interlamination of sand, silt, and clay point to sedimentation from gentle currents of pulsating velocity which alternatively caused sand ripples to migrate and then halted to allow clay to settle out of suspension. The resultant bedding type and the associated intensive burrowing are closely comparable to Recent tidal flat deposits such as those described from the North Sea coasts (Van Straaten, 1954; Evans, 1965). Likewise, the morphology and sediments of the channels are analogous to those formed by the tidal gullies which drain Recent mud flats (e.g. Van Straaten, 1954, Fig. 4). Further evidence of the shallow environment of this facies is provided by the lignites and rootlet horizons which may have originated in salt marshes on the landward side of the tidal flats. Rare desiccation cracks also testify to the shallow origin of this facies.

## Cross-bedded sand and shale facies

*Description*

This is the most southerly of the four facies belts which are aligned subparallel to the Sirte basin shore. It is best developed in the south scarp of the Jebel Zelten and also occurs, interbedded with the previously described facies, at the base of jebels exposed northwards as far as the Marada escarpment.

This facies is divisible into three interbedded sub-facies. Two-thirds of it is composed of poorly sorted pale yellow unconsolidated sands. These range in grain size from coarse to fine. The coarser sands occur in sequences 5–6 m thick with erosional channelled bases, often veneered by thin quartz pebble conglomerates. Internally they are both tabular planar and trough cross-bedded with set heights of 30–40 cm arranged in vertically grouped cosets. The orientations of foresets and trough axes indicate deposition from unidirectional northerly flowing currents (Fig. 7.2). The finer sands are argillaceous, massive, and, sometimes, flat-bedded.

The second sub-facies, interbedded with the first, consists of laminated shales which occur both as sheets and infilling abandoned channels. They can be distinguished from the laminated shale facies to the north since they are not calcareous and swell and fall apart in water. This phenomenon suggests that they are largely composed of montmorillonite.

The third sub-facies consists of thin sequences, only a few centimetres thick, of lignite, sphaerosideritic limestone, and ferruginized sand pierced throughout by rootlets.

The cross-bedded sand and shale facies contains an abundant, diverse, and well-preserved vertebrate fauna. This includes bones of terrestrial mammals such as ancestral elephants, camels, giraffes, antelopes, and carnivores, together with aquatic forms such as crocodiles, turtles, and fish. These bones occur within the sands together with transported tree trunks. The interbedded shales contain plant debris and the continental gastropod *Hydrobia*.

*Interpretation*

This facies differs from the three previously described in the dominance of terrigenous sediment and continental fossils. These facts, together with the sedimentary structures and northerly directed palaeocurrents, suggest deposition in a fluviatile environment. (See Chapter 2 for the diagnostic features of alluvium.)

Accordingly, the sand sub-facies can be attributed to deposition within river channels, while the shales and fine sands probably originated on levees and floodplains. The channel-fill shales are abandoned ox-bow lake deposits. The thin limestones, ferruginous layers, lignites, and rootlet beds represent old soil horizons which perhaps formed on the levees and low-lying flood basins between channels. The large amounts of shale imply that the alluvium was due to sinuous meandering rivers rather than braided ones. The vertebrate fauna suggests a savannah climate.

In conclusion, it can be seen that the cross-bedded sand and shale facies originated in a low-lying alluvial coastal plain which, seawards, merged imperceptibly into the tidal flats to the north.

## Calcareous sandstone channel facies

*Description*

The four facies belts paralleling the Sirte basin just described are locally cross-cut by northerly trending calcareous sandstone channels of the fifth and last facies.

These can arbitrarily be sub-divided into two types: small isolated channels of sandy limestone, and large radiating channel complexes of calcareous sandstone.

Channels of the first type are concentrated near the base of the southern scarp of the Jebel Zelten, where they are interbedded with the fluviatile deposits. These also occur at the base of jebels around the Zelten oil field and can be traced at the same level as far north as the Marada scarp. Channels of this type are about 10 m deep and 300 m wide. In some areas they are well-exposed where they have been exhumed from the softer sands and shales with which they are interbedded (Fig. 7.9). These channels are composed of poorly sorted medium- and coarse-grained sandy limestones with fragments of marine shells mixed with quartz and micrite matrix. They are floored by intraformational bored limestone pebble conglomerates. The sands are cross-bedded in a wide variety of scales and types, and, less commonly, massive, flat-bedded, or rippled. Burrowing is present, especially in the finer sediment towards the top of each channel which is often intensively burrowed and sometimes shot through with plant rootlets.

The major channels occur at two points and, due to the good exposure, can be isopached (Fig. 7.1). At Reguba this facies is about 200 m thick. Near the crest of the south scarp of the Jebel Zelten is a sheet of calcareous sandstone, lenticular in an east-west direction, about 25 km wide and 30 m deep. On the north scarp of the Jebel Zelten this has split up into a series of discrete channels

*Figure 7.9*  Meandering estuarine channel of resistant sandy limestone, exhumed from soft fluvial sands and shales, now crops out as a long sinuous flat-topped jebel. From Selley, 1968, Plate 24(a), by courtesy of the Geological Society of London.

which can be traced north to the upper part of the sections in jebels around the Zelten oil field. One or two sandy channels occur at the same level in the Marada scarp.

Petrographically this facies consists of coarse and very coarse, sometimes pebbly, calcareous sandstones. These are arranged in a series of coalesced channels infilled with various kinds of large-scale trough and planar cross-bedding. These sometimes show penecontemporaneous deformation attributable to quicksand movement. Apart from bioturbation, this facies contains no fossils *in situ*. There are fragments of marine shellfish, bones, teeth, and wood.

Palaeocurrents determined from cross-bedding are bipolar with the major mode pointing northwards (Fig. 7.2). Around the Jebel Zelten palaeocurrents plotted at outcrop show a regionally radiating pattern.

### Interpretation
Clearly this facies was deposited from fast-flowing currents confined to channels. The palaeocurrents and the mixture of marine carbonate sediment and shells with terrigenous sand, bones, and wood indicate to and fro current movement. It seems highly probable, therefore, that this facies originated in estuaries subject to strong tidal currents.

The location of the two major channel complexes at Reguba and Jebel Zelten may not be a matter of chance. They both trend along north-south pre-Miocene palaeohighs which host several major oil fields. This suggests that the negative movement occurred along these two trends in Miocene time, favouring the development of estuaries where they cross-cut the shoreline.

## General discussion of the Miocene shoreline of the Sirte basin

In conclusion, it can be seen that this case history provides a good example of mixed carbonate:terrigenous shoreline. From north to south relatively deep water carbonates pass up slope into coarser cross-bedded shell sands, deposited on shoals and bars. These inter-finger southwards with fine-grained terrigenous muds and sands of lagoonal and tidal flat origin. In turn these pass landwards into an alluvial coastal plain facies. Locally, this shoreline was interrupted by two major estuaries which supplied coarse sand to the shoreline (Fig. 7.10).

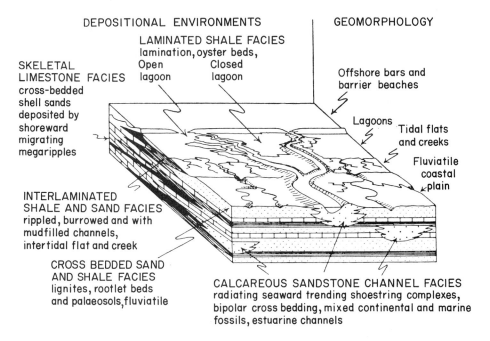

*Figure 7.10* Block diagram illustrating the supposed origin of the Miocene shoreline of the Sirte basin, Libya. From Selley, 1968, Fig. 17, by courtesy of the Geological Society of London.

Currents were seldom strong enough to carry this detritus out to the bar zone. Subtle syn-sedimentary movement seems to have controlled the distribution of facies. The concentration of coarse grain size and cross-bedding in the limestones over the east-west trending crest of the Zelten anticline suggests that it may have been a Miocene palaeohigh on which the barrier carbonates were deposited. Likewise the two major estuaries seem to have been located along northerly subsiding axes.

Palaeocurrent analysis shows that the carbonate sands were deposited by predominantly up slope shoreward-flowing currents, thus coming to rest in waters shallower than that in which they formed. In contrast, quartz sand was carried off the Sahara shield to be deposited in the alluvial plain and tidal flats. Only the finest fraction was transported as far as the lagoons. Mixing of the land-derived quartz sand and marine carbonate detritus occurred only in the estuarine channels, perhaps due to tidal currents.

## GENERAL DISCUSSION AND ECONOMIC ASPECTS

Shorelines where barrier carbonates are juxtaposed with continental terrigenous facies are transitional between clastic shores where carbonate sedimentation, if present, is restricted to the offshore zone, and carbonate shores where terrigenous sediment is negligible. These are described in the previous and subsequent chapters respectively.

Apart from the Libyan Miocene case, other mixed shorelines occur in the Permian rocks of West Texas (Chapter 9). Here shoal calcarenites pass landwards through lagoonal deposits into continental red beds and evaporites. These are overlain by reef limestones with similar shoreward facies changes.

Indeed, the combinations of conditions which favour the occurrence of mixed carbonate:clastic shores (i.e. low influx of terrigenous sediment and aridity) are particularly favourable for reef growth. Perhaps reefs are more characteristic of mixed shorelines than are carbonate sand banks.

These shorelines are of considerable economic significance. This need not be discussed here. The importance of alluvial deposits has already been described on pages 66 and 161. The economic aspects of carbonate bars and reefs are discussed on pages 204 and 234 respectively.

## REFERENCES

The account of the Libyan Miocene shoreline was based on:

Doust, H., 1968. *Palaeoenvironmental Studies in the Miocene* (*Libya, Australia*). Vol. I. Unpublished Ph.D. Thesis. University of London. p. 254.

Savage, R. J. G., and White, M. E., 1965. Two mammal faunas from the early Tertiary of Central Libya. *Proc. Geol. Soc. Lond.* No. 1623, pp. 89–91.

Selley, R. C., 1966. The Miocene rocks of Marada and the Jebel Zelten: a study of shoreline sedimentation. *Petrol. Explor. Soc. Libya.* p. 30.

——, 1967. Paleocurrents and sediment transport in the Sirte basin, Libya. *J. Geol.*, **75**, 215–23.

——, 1968. Facies profile and other new methods of graphic data presentation: application in a quantitative study of Libyan Tertiary shoreline deposits. *J. Sediment. Petrol.*, **38**, 363–72.

——, 1968. Nearshore marine and continental sediments of the Sirte basin, Libya. *Quart. Jl. Geol. Soc. Lond.*, **124**, 419–60.

——, 1972. Structural control of Miocene sedimentation in the Sirte basin. In: *The Geology of Libya.* (Ed. C. Gray) The University of Libya. 99–106.

Other references listed in this chapter were:

Evans, G., 1965. Intertidal flat sediments and their environments of deposition in the Wash. *Quart. Jl. Geol. Soc. Lond.*, **121**, 209–45.

Ginsburg, R. N., 1953. Beachrock in south Florida. *J. Sediment. Petrol.*, **23**, 85–92.

Jindrich, V., 1969. Recent carbonate sedimentation by tidal channels in the Lower Florida keys. *J. Sediment Petrol.*, **39**, 531–53.

McKee, E. D., and Sterrett, T. S., 1961. Laboratory experiments on form and structure of longshore bars and beaches. In: *Geometry of Sandstone Bodies*. (Ed. J. A. Peterson and J. C. Osmond) Amer. Assoc. Petrol. Geol., pp. 13–28.

Van Straaten, L. M. J. U., 1953. Composition and structure of Recent marine sediments in the Netherlands. *Leid. Geol. Meded.*, **19**, 1–110.

# 8 Shelf deposits: carbonate and terrigenous

## INTRODUCTION: A GENERAL THEORY OF SHELF SEA SEDIMENTATION

The preceding three chapters described shoreline sediments. Where sediment input from the land is so great that marine currents cannot re-distribute it, deltas form. Where marine currents are competent to re-distribute the sediment linear bar and barrier coasts form.

With declining sediment input lowlying tidal flats may pass seawards into marine shelves where offshore bars or tidal current sand-waves migrate across scoured erosional surfaces. Where terrigenous sediment input to the shoreline is negligible they may be replaced by carbonate sands and muds. This chapter is concerned with both the terrigenous and the carbonate deposits of shelf seas.

There are many areas in the world where relatively thin (generally less than 1000 m) carbonate sequences cover thousands of square miles with easily correlatable layer-cake stratigraphies. Palaeontology indicates that these deposits were formed in relatively shallow open marine conditions. These deposits, though they may at present occupy tectonic basins, must have been deposited on vast shelves with only the gentlest of bottom slopes.

It is difficult to use modern shelf seas as analogues for their ancient counterparts for two reasons. First, at the present time the earth seems to lack the vast subhorizontal shelf areas that existed in earlier times. Secondly, present-day shelves have recently been subaerially exposed during glacially induced drops in sea level. The modern sediments of marine shelves, though now reworked by tidal currents, were actually transported by fluvio-glacial processes. Thus these relict sediments are not suitable analogues for ancient shelf deposits. (Emery, 1968; McManus, 1975). One or two modern shelves are known, however, in which the sediment is graded and in equilibrium with the present-day current regime. The Bering shelf has been cited as one such example (Sharma, 1972). For further accounts of modern marine shelves see Swift *et al.* (1973), and Stanley and Swift (1976). One of the best known of modern carbonate shelf seas is the Arabian Gulf (Purser, 1973).

A general theory of carbonate shelf sedimentation was proposed by Irwin (1965) based on a study of Upper Palaeozoic carbonates of the Williston basin, North America. This case history is described in detail in this chapter. First Irwin's epeiric sea model will be described and discussed.

Within limestone shelf sequences, such as those of the Williston basin, it is possible to define three major sedimentary facies which pass laterally into one another from the centre of the basin to its margin in the following manner:

(1) *Calcilutites:* grading via skeletal wackestones and packstones into
(2) *Grainstones* (skeletal and oolitic): grading via packstones into
(3) *Pelletoidal wackestones, microcrystalline dolomites and evaporites.*

Irwin explained these facies relationships thus: If one considers a marine shelf, it consists of two parallel horizonal surfaces of sea level and wave base. These transect the sloping sea floor shorewards (Fig. 8.1). In the deeper part of the basin, below wave base, fine-grained laminated mud will settle out of suspension. The fauna will generally be preserved *in situ* and unfragmented. These conditions may extend over thousands of square miles. Irwin termed this the 'X' zone. As the sea floor rises towards the shelf margin it becomes influenced by two factors: light and turbulence. The lower limit of the photic zone is at about 8 cm, though it varies with turbidity. Photosynthesis by algae can only take place within the photic zone. This process has two important consequences. Photosynthesis increases the oxygen content of water, thus making it inhabitable for animal life. The algae also provide food for animals. Thus within the photic zone phyto- and zooplankton thrive and form the basis of the food chains of larger organisms. Many of these secrete a calcium carbonate skeleton. Where the sea floor lies within the photic zone large-shelled organisms can colonize it. The shallow sea floor is also generally exposed to some degree of current activity. This tends to destroy skeletal material, generating bioclastic sands and muds. There is no clearly defined lower limit to current activity. On modern ocean floors currents are sometimes sufficiently powerful to transport sand (see p. 253). A very old geological concept is the one of 'effective wave base'. This was taken to mean the lower limit of sediment transport due to the orbital motion

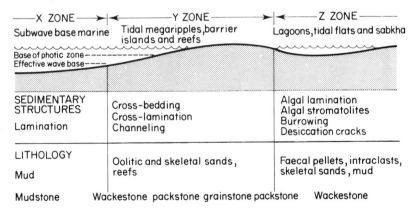

*Figure 8.1* Cartoon to demonstrate the X-Y-Z zone model of shelf sedimentation proposed in Irwin (1965).

of waves, and was believed to extend down to about 200 m, the edge of the continental shelf. The lower limit of wave action is controlled by wavelength and periodicity. These of course vary from place to place, and seasonally. Furthermore continental shelves are affected not only by wind-generated waves, but also by tidal currents. It is therefore very difficult to put an exact depth to the lower limit of significant current action.

Returning to Irwins' X-Y-Z zone model we note that moving from basin centre to margin there is a critical depth at which current activity on the sea bed is sufficient to winnow out mud, and allow only sand to be deposited. We may for the sake of convenience refer to this depth as 'effective wave base', but if we do then we imply no exact depth, and we also allow that this surface is not just the lower limit of wave-induced currents, but also of tidal ones.

Irwin designated this shallow high energy environment the 'Y zone'. The sea floor may be a scoured bed rock surface where conditions are very turbulent, or an area of sand, where they are less extreme. On modern continental shelves sand waves exist which are similar in many ways to the eolian dunes of deserts, see for example Bouma *et al.* (1982) and Stride (1982).

These sands may be terrigenous, but in areas where there is little land derived detritus, the sea floor above the photic zone may be colonized by lime-secreting organisms. Current action will constantly smash up the skeletal material. The resultant bioclastic sand may be transported in migrating ripples and megaripples, while the finer mud may be carried out in suspension to be finally deposited on the deeper quieter floor of the X zone. In extreme cases the sea floor may be colonized by colonial organisms whose skeletons are so resistant to current action that reefs develop. Reefs are a particular type of Y zone environment of such importance that they will be considered separately in the next chapter.

In the Y zone ooids sometimes tend to develop. These are spherical lime grains formed by the growth of concentric layers of aragonite needles around a nucleus, generally a skeletal grain or quartz particle. Ooid formation tends to occur where the warm waters of a sheltered shelf or lagoon mix with cooler open sea water. Thus cross-bedded oolitic sands tend to be deposited in the Y zone either in restricted tidal channel bodies or in more extensive blankets where ooid megaripples migrated across the shelf. The edge of the modern Bahama Bank is a famous recent example of this situation (Illing, 1954; Newell, Purdy and Imbrie, 1969; Purdie, 1963; Ball, 1967; and Enos, 1982).

Thus the Y zone environment deposits skeletal or oolitic grainstones in its most turbulent parts and packstones in more sheltered areas. These grade into wackestones as the Y zone passes into the deeper waters of the X zone. Whereas the X zone may cover thousands of square kilometres, the Y zone is often a belt only a few tens of kilometres wide extending parallel to the shelf margin. Differential subsidence and uplift often complicates this simple pattern.

Shoreward of these high-energy barrier sands the sea bed continues its gentle landward climb. In these restricted low-energy waters lagoonal conditions

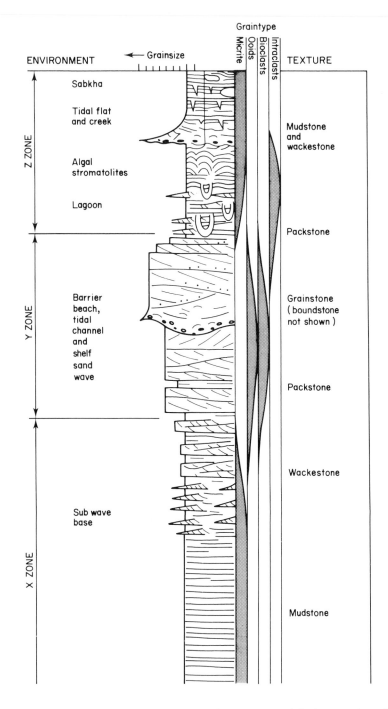

*Figure 8.2* Sketch to illustrate the vertical sequence of facies produced by the basinward progradation of a single increment of X, Y and Z zone environments.

prevail. Recent carbonate lagoons are well-documented in the references cited above (see also Shinn, 1982; and p. 171). They are characterized by skeletal and faecal pellet sands which, since they are deposited in quieter conditions than those of the barrier, are micritic (packstones and wackestones). These grade landwards into tidal flats of laminated, sometimes burrowed, carbonate muds. The lamination is often algal in origin, and algal stromatolites may occur. The tidal flats are drained by tidal creeks. Subaerial exposure favours the early lithifaction of intertidal sediment. Subsequent submergence favours the penecontemporaneous erosion of newly cemented lime sediment. Thus intra-formational clasts (intraclasts for short) are common in lagoonal and tidal flat sediments. They may occur as conglomerates on the floors of tidal channels, or mixed in with skeletal or faecal pellet muds. If salinities are high, dolomites and evaporites may form. There has been considerable speculation whether these are precipitated directly on the floor of lagoons or whether they are due to penecontemporaneous replacement of carbonate mud within tidal flats. In Recent arid shorelines, such as those of the Baja California and the Trucial Coast of the Arabian Gulf, wide salt flats (sabkhas) are developed. In these, evaporite minerals form at the present time (Shearman, 1963 and 1966; Holser, 1966; Evans *et al.*, 1969; Kinsman, 1969). It seems that, as the sun beats down, capillarity draws lagoonal brines into the pore spaces of the sabkha carbonates. Here, as the fluids evaporate and concentrate, they replace the host sediment to form dolomite, gypsum, anhydrite, halite, and other evaporite minerals. This facies of pelletal limestones, dolomites, and evaporites Irwin termed the 'Z' zone. Figure 8.2 shows the vertical sequence of carbonate facies deposited by the progradation of an X, Y, Z zone sequence of environments.

This then is the interpretation which may be put forward as a general explanation of carbonate shelf shoreline sedimentation. This is a very simple model, and other more sophisticated schemes have been proposed which take account not only of wave base, but also of the depth of the photic zone, and of the ecologic zonation of lime-secreting organisms (Lees, 1973; Wilson, 1975).

Mississippian deposits of the North American Williston basin will now be described and discussed since they are a typical example of carbonate shelf sediments.

## MISSISSIPPIAN (LOWER CARBONIFEROUS) DEPOSITS OF THE WILLISTON BASIN, NORTH AMERICA: DESCRIPTION

The Williston basin covers many thousands of square miles of the states of Montana and the Dakotas of the USA and the provinces of Alberta, Saskatchewan, and Manitoba in Canada. It contains over 5 km of sediments of all geological periods. These rocks are largely of shallow-water origin and have suffered only the gentlest tectonic deformation, basinward dips being locally interrupted by anticlinal flexures. In Upper Palaeozoic time carbonate

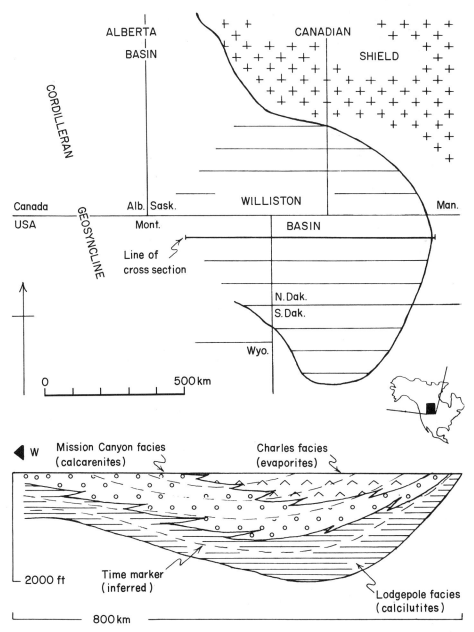

*Figure 8.3* Upper: map of the Mississippian (Lower Carboniferous) deposits of the Williston basin. Modified from Carlson and Anderson, 1965, Fig. 1. From the *Bulletin of the American Association of Petroleum Geologists* by courtesy of the American Association of Petroleum Geologists. Lower: cross-section of Mississippian (Lower Carboniferous) deposits of the Williston basin. Modified from Smith *et al*. 1958, Fig. 3, by courtesy of the American Association of Petroleum Geologists.

sedimentation took place from the Williston basin northwards through the Alberta and Mackenzie basins to the Arctic. These deposits consist largely of shales, limestones (sometimes reefal), dolomites, and evaporites. Oil-bearing Devonian reefs are described in the next chapter. Petroleum is also present in Mississippian carbonates of the southeastern edge of the Williston basin in Canada and northern USA. Intensive study of these beds led to the concept of X, Y, and Z zones stated in the previous section. These rocks will now be described.

The Mississippian section of the Williston basin is over 700 m thick with a depocentre in the western part of North Dakota. Stratigraphical nomenclature is complex and regionally variable. For the purposes of this account the system proposed by Carlson and Anderson (1956, p. 1340) is used (Fig. 8.3). A series of stratigraphic intervals are defined by thin persistent evaporites and clastics which can be correlated regionally in sub-surface by gamma-ray peaks on well logs. The marker horizons which define these intervals are believed to approximate to time planes. Three facies diachronously cross-cut these from top to bottom as follows:

(*i*) *Charles facies:* cyclic evaporites.
(*ii*) *Mission Canyon facies:* bioclastic limestones, dolomites, and oolites.
(*iii*) *Lodgepole facies:* thin-bedded argillaceous limestones.

These three facies interfinger with one another laterally and over-step each other laterally towards the centre of the basin. They will now be described in turn:

## Charles facies

These beds are developed around the southern and eastern edge of the Williston basin and extend diachronously basinward over the underlying Mission Canyon facies. Lithologically this facies consists largely of dolomite and anhydrite with minor amounts of halite, shale, and sandstones. These are rhythmically arranged with the following general sequence: each rhythm begins with pelmicrites and biomicrites which grade up into microcrystalline dolomites with scattered shell fragments. These contain anhydrite stringers and nodules towards the top, grading in turn up into anhydrite rocks with dolomite veins.

## Mission Canyon facies

The second facies is overlain by, and passes up slope into, the Charles facies, and overlies and passes basinwards into the Lodgepole facies. Petrographically these beds are well-sorted, lime mud free calcarenites, sometimes dolomitized or sparite-cemented, but often retaining primary interparticle porosity. These rocks are sometimes oolites and sometimes skeletal sands, often composed largely of crinoid debris. Up slope they grade into pelsparites which, with increasing lime mud matrix, pass into the pelmicrites of the Charles facies. The abundant but often fragmented fossils of the Mission Canyon facies include crinoids, brachiopods, bryozoa, corals, foraminifera, and algae.

## Lodgepole facies

This facies is best developed in the central part of the basin where it is overlain by the Mission Canyon sediments. Lithologically these rocks are thin-bedded or laminated dark grey argillaceous limestones. They are locally siliceous and interbedded with cherts. The fauna is similar to that of the Mission Canyon facies but much less abundant, better preserved, and with few corals or algae. Fossils are sometimes silicified.

# MISSISSIPPIAN (LOWER CARBONIFEROUS) SEDIMENTS OF THE WILLISTON BASIN, NORTH AMERICA: INTERPRETATION

It should be clear from the introductory section of this chapter that the Mississippian deposits of the Williston basin are of carbonate shelf type. The Lodgepole, Mission Canyon, and Charles facies correspond to the X, Y, and Z zones defined by Irwin (1965).

Considered in more detail the fine-grain size and fauna of the Lodgepole facies indicate sedimentation in a low-energy marine environment. The large lateral extent of this facies shows that deposition was in an open sea with sedimentation below wave base, and/or away from bottom currents and terrestrial detritus.

The fragmental fauna, oolites, and clean-washed texture of the Mission Canyon facies indicate deposition in a high-energy marine environment with winnowing of skeletal sands on migrating shoals and banks. The pelletoidal micrites, crystalline dolomites, and evaporites of the Charles facies suggest deposition in a restricted marine environment cut off from the open sea by bars of skeletal sands. Pelmicrites are typical of Recent lagoons (p. 188). The microcrystalline dolomites and evaporites may have formed by primary precipitation on lagoon floors or by diagenesis in intertidal and supratidal flat deposits analogous to Recent sabkhas as already described (p. 188).

From the stratigraphic relations of the three facies it is apparent that they developed synchronously and migrated basinward one on top of the other. Cyclicity in the evaporite facies may reflect minor fluctuations of the shoreline superimposed on this general basinward regression.

It can be seen therefore that the Mississippian rocks of the Williston basin are a good example of an ancient marine carbonate shelf: shoreline environment.

# DISCUSSION OF CARBONATE SHELF SEDIMENTS

Carbonate sediments have been studied in much greater detail since Irwins' X-Y-Z zone model was produced in the 1960s. More sophisticated sedimentary

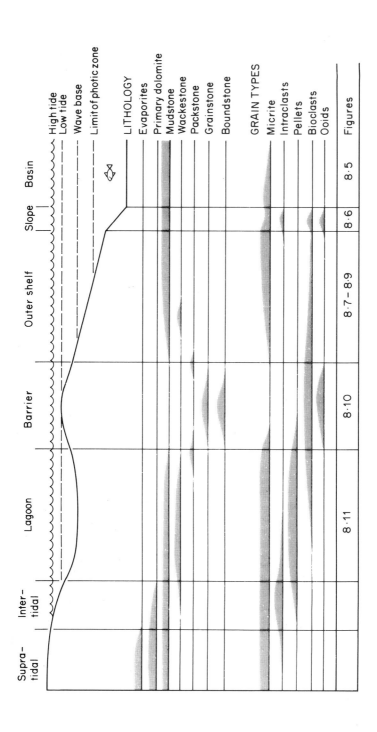

*Figure 8.4* Complex carbonate shelf model.

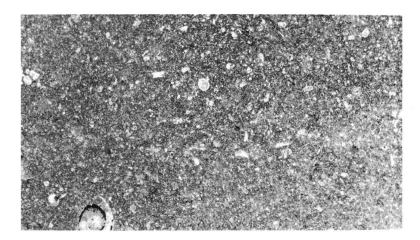

*Figure 8.5* Photomicrograph of Palaeocene carbonate micro-facies from the sub-surface of the Sirte basin, Libya. Calcilulite with pelagic foraminifera, basinal environment. Released by courtesy of Oasis Oil Company of Libya Inc. Field of view 8 mm wide.

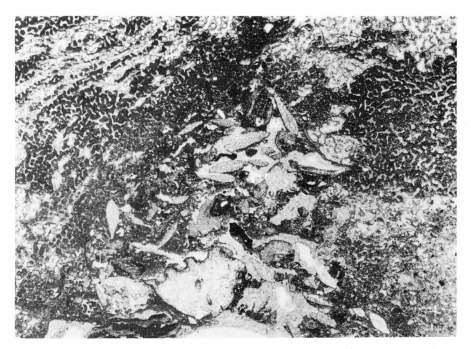

*Figure 8.6* Thin section of Montpelier Formation (Miocene), Jamaica. This is an allodapic wackestone deposited in a deep water trough. It contains a derived shallow water fauna that includes large benthonic foraminifera and Rudist fragments.

models have been developed by Lees (1973), Wilson (1975), Reeckmann and Friedman (1982), and Read (1985). These studies define up to nine distinct environments and resultant facies.

Figure 8.4 illustrates a more complex carbonate shelf model based on these four authors. There are still three major environments, termed basin, outer shelf, and inner shelf, but these only partially correlate with Irwins' X, Y and Z zones.

The basin environment extends beyond the continental slope. It passes with increasing water depth below the carbonate compensation depth at which skeletal debris dissolves. This thus merges into the abyssal environment that is dealt with in Chapter 11. Basinal carbonates are fine-grained micrites with a pelagic fauna of radiolaria and the smaller pelagic foraminifera, such as Globigerinids (Fig. 8.5).

There is commonly a continental margin at the boundary between the basin and outer shelf. This margin may form a distinctive slope environment on which carbonate debris flows and turbidites are deposited. The resultant sediments are composed of fragmented derived shallow marine detritus, and are often referred to as allodapic limestones (Meischner, 1965). Figure 8.6 illustrates an example of this type of limestone from Jamaica. The slope environment passes landward into the outer shelf environment. Where the outer shelf is below effective wave base fine-grained lime mudstones will be deposited. These are broadly similar to the basinal deposits previously described. A good example of ancient outer shelf deposits occur in the Cretaceous rocks of the Middle East,

*Figure 8.7* Photograph of the Upper Cretaceous chalk cliffs of Beer, Devon showing the regular bedding of this X zone shelf deposit.

North Africa, Northwest Europe, and the USA. These include the Austin Chalk of Texas and the Chalk of England (Fig. 8.7). Chalk is a fine-grained lime mudstone composed largely of the skeletons of a group of blue-green algae termed the nannoplankton. As their name suggests, these have a pelagic lifestyle. When they die their skeletons, termed coccospheres, sink to the sea bed, sometimes disaggregating into their constituent plates, termed coccoliths (Fig. 8.8). Chalk limestones also contain calcispheres (microscopic balls of unknown origin), small pelagic Globigerinid foraminifera, and a sparse macrofauna. This includes echinoids, brachiopods, lamellibranchs, sponges, bryozoa and solitary corals. Chalk is often extensively bioturbated, with the ichnogenera *Thalassinoides* and *Zoophycus* (Bromley, 1967). Chert nodules and

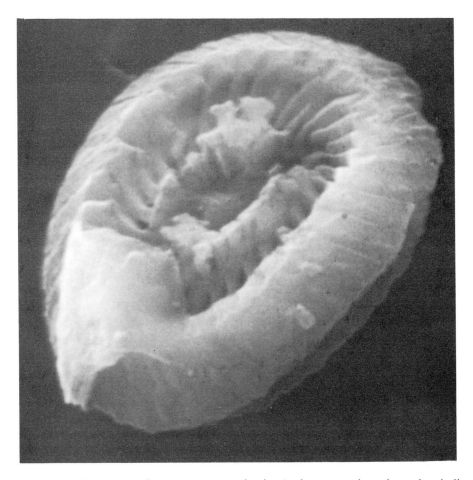

*Figure 8.8a* Scanning electron micrograph of a single coccosphere from the chalk showing constituent coccoliths. Lower Chalk, Folkestone. ×11 000. By courtesy of J. Young.

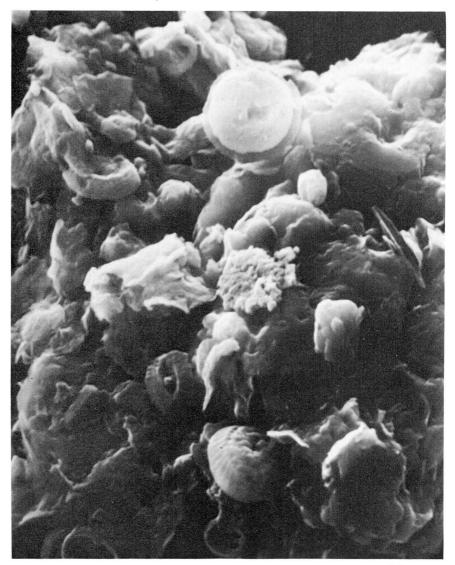

*Figure 8.8b.*  Scanning electron micrograph of chalk showing the coccolithic nature of the deposit. Lower Chalk, Folkestone. ×2 500. By courtesy of J. Young.

beds are common, known by the natives of southern England as 'flint'. Chert often preferentially replaces burrows (Fig. 8.9). The uniform depositional nature of the Chalk sea is demonstrated by the existence of laterally persistant 'hardground' horizons. These are intraformational erosion surfaces that are commonly bored and phosphatized. They are encrusted by the shells of beasts that grew on a rocky substrate. Each erosion surface is overlain by a thin layer

*Figure 8.9* Photograph of chalk showing preferential chertification of burrows. Annis's Knob. Beer. Devon.

of argillaceous chalk with an intraformational basal conglomerate of chalk clasts. Individual 'hardground' horizons can be traced for hundreds of kilometres across Northwest Europe (Jefferies, 1963).

The fine grain size and fauna of the Cretaceous Chalk indicate deposition in a low energy marine environment. The absence of benthonic algae imply that the sea floor was below the limit of the photic zone. Detailed palaeontological studies suggest water depths of between 200–600 m (Hakansonn *et al.*, 1974). The Cretaceous chalks thus provide a good example of an outer shelf environment that developed when there was a global marine transgression across the continental shelves. Locally, however, rifts developed across the shelves. These troughs were infilled with chalk reworked from the shelves and deposited from turbidity currents and debris flows. An example of this is known in the Central Graben of the North Sea (Kennedy, 1980).

The outer and inner shelves are commonly separated by a high energy barrier, correlative with Irwins' Y zone. As already discussed this is an area of carbonate formation, and results in the deposition of banks of skeletal and oolitic sands, or in reef formation. Figure 8.10 illustrates an example of a shoal grainstone. The reef environment is such an important one that it is dealt with in great detail in the next chapter.

Figure 8.4 shows that the inner shelf (Irwins' Z zone), can be divided into subtidal, intertidal and supratidal subenvironments. The subdidal environment may be such a broad region that it merits the term 'inner shelf', or so narrow

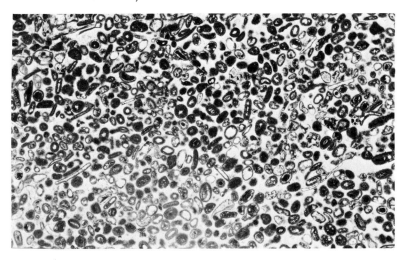

*Figure 8.10* Photomicrograph of Palaeocene carbonate micro-facies from the sub-surface of the Sirte basin, Libya. Grainstone of ooids and bioclasts, high-energy shelf environment. Released by courtesy of Oasis Oil Company of Libya Inc. Field of view 6 mm wide.

that it is more properly a lagoon. These sheltered shallow environments permit considerable biological carbonate formation, but supply of nutrients is less than along the barrier and growth correspondingly diminished. The resultant sediments tend to be skeletal wackestones. The fauna is generally unfragmented and sometimes in growth position. Whereas the outer shelf environment is dominated by small pelagic foraminifera the inner shelf is characterized by larger benthonic ones. Inner shelf sediments also contain large amounts of faecal pellet material (Fig. 8.11). Early diagenesis and penecontemporaneous erosion of faecal pellet sands forms botryoidal composite grains termed 'grapestones'. Intertidal carbonate sediments are similar to subtidal ones in many ways. They may, however, contain algal stromatolites ranging from gently undulose laminations to more regularly organized Collenia colonies. Subaerial exposure of the intertidal zone coupled with intermittant flushing by freshwater give these deposits characteristics that are absent in their subtidal counterparts. These include desiccation cracks and occasional rootlets due to marsh vegetation. A particularly characteristic feature of intertidal carbonate sediments is the presence of irregular pore systems that parallel bedding. These are variously described as 'fenestral porosity' (Tebbutt *et al.*, 1965), 'birds-eye' structure, or loferite (Fischer, 1964). This feature has variously been attributed to the buckling of sediment laminae during subaerial exposure, to deformation due to the escape of biogenic methane, and to the leaching of organic algal mucilage. Early lithifaction and penecontemporaneous erosion often gives rise to the formation of intraclasts. These may be so abundant as to form intraformational

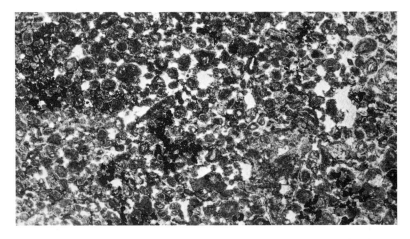

*Figure 8.11*   Photomicrograph of Palaeocene carbonate micro-facies from the sub-surface of the Sirte basin, Libya. Fenestral limestone of lime mud pellets with sparse lime mud matrix. Uncompacted state of pellets suggests early diagenesis possibly due to penecontemporaneous sub-aerial (?tidal) exposure. Low-energy inner shelf. Field of view 7 mm wide.

conglomerates in tidal channels, or they may become mixed in with more open inner shelf sediments. The intertidal environment passes landwards into the supratidal one (Fig. 8.4). This is basically similar to the inner shelf deposits, but the evidence of subaerial exposure is more intense, sometimes including palaeokarstic solution surfaces. Similarly there is commonly more evidence of early diagenesis. In arid climates 'sabkha' salt marsh conditions may prevail, with penecontemporaneous dolomitization and evaporite mineral growth. In areas of stable shelf sedimentation there are often genetic increments, only a few metres thick, that span the spectrum from outer shelf to intertidal deposits. In areas of tectonic activity, with well defined shelf margins, the facies belts are better defined and may be mapped.

## TERRIGENOUS SHELF DEPOSITS

It is possible to extend this concept of high- and low-energy shelf environments outside the realm of carbonates to terrigenous sediments.

The Lower Palaeozoic rocks of the Saharan and Arabian Shields show a broad sequence of three sedimentary facies (see Bender, 1968, 1975; Helal, 1965; Klitzsch, 1966; Bennacef *et al.*, 1971).

(1) *Graptolite bearing shales and fine sandstones.* These attain their maximum development in basins, such as the Murzuk and Kufra basins of Libya and the Tabuk basin of Arabia. This facies is generally of Ordovician to Silurian age.

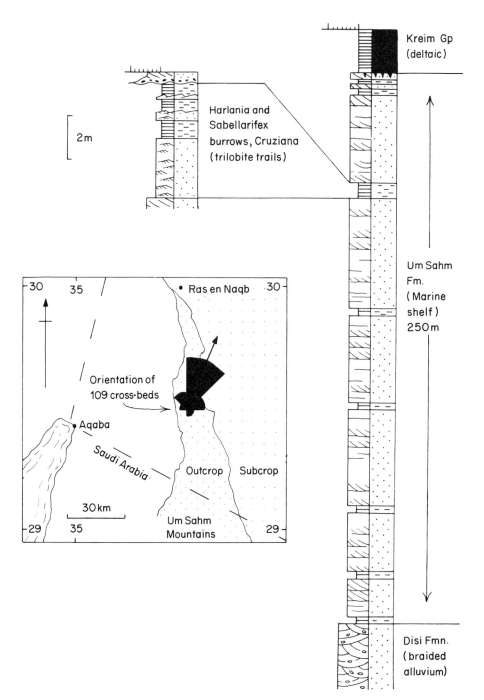

Figure 8.12 Summary section and outcrop map of the Um Sahm Formation, Jordan. The major part of the sequence is composed of cross-bedded shoal sands (Y zone), with minor bioturbated upwarding-fining tidal flat shales (Z zone) as shown inset.

(2) *Medium to fine well-sorted sands* of wide lateral extent. These contain thin shales with *Cruziana* (trilobite tracks), and the burrows *Tigillites* ( = *Scolithos* = *Sabellarifex*) and *Harlania* ( = *Arthrophycus*). Examples of this facies are the Um Sahm Formation of Jordan and the Haouaz Sandstone of southern Libya (Ordovician).

(3) *Coarse pebbly cross-bedded sands* of wide lateral extent, generally unfossiliferous, e.g. the Saq Sandstone of Saudi Arabia, the Saleb, Ishrin, and Disi Formations of Jordan, and the Hassaouna Sandstone of southern Libya (Cambro-Ordovician).

PreCambrian igneous and metamorphic basement complex.

*Figure 8.13* Outcrop of Um Sahm Formation in the Southern Desert, Jordan. Laterally continuous ledges due to soft weathering shale bands with trace fossils.

Sedimentological and trace fossil studies of the above sequences in the Southern Desert of Jordan suggest that the three facies are, from bottom to top, braided alluvial, marine shelf, and deltaic deposits (Selley, 1970; 1972). The supposed marine shelf facies, the Um Sahm Formation of Jordan, will now be described (Fig. 8.12).

Some 250 m thick, this formation crops out widely over the southern desert of Jordan and can be traced eastwards into Saudi Arabia (Helal, 1965). Its southwestward limit is erosional. Northeastwards it is overlain by the graptolitic deltaic shales and sands. The Um Sahm Formation is composed of fine- to medium-grained well-sorted protoquartzites which are dirty white when fresh but weather to a dark brown (Fig. 8.13). The sands are massive or tabular-planar cross-bedded with sets 5–15 cm high. These are grouped in cosets of considerable lateral extent (Fig. 8.14). Unlike the braided alluvial facies beneath (p. 64) they do not occur in channels. Foreset orientations indicate deposition by currents which flowed northeastwards off the Arabian Shield. At several levels there are layers, only a metre or so thick, of laminated grey siltstone and micro-crosslaminated very fine sand. These sequences have sheet geometries and can be traced for kilometres along and perpendicular to the palaeoslope.

Trace fossils in the shales include *Cruziana* tracks, attributed to trilobites, vertical burrows named *Sabellarifex* ( = *Scolithos* = *Tigillites*) and sub-horizontal burrows of the type *Harlania* ( = *Arthrophycus*).

The cross-bedding and sheet geometry of the Um Sahm sands suggest deposition from megaripples by unidirectional open-flow (i.e. not channel-confined) currents. The shales and rippled sands indicate intermittent low-energy conditions. If the *Cruzianas* are indeed trilobite trails then these waters were probably marine. It seems most probable that the Um Sahm sands were deposited

*Figure 8.14*   Coset of tabular planar cross-bedding in Um Sahm Formation, Jordan.

from migrating shoals on an open marine shelf and are therefore a clastic example of Irwin's Y zone environment. The shale units however could be due either to the low-energy offshore sub-wave base X zone, or to the low-energy shallow Z zone to the lee of the sand shoals. The clue to this dilemma is provided by the burrows and trails. Studies of Recent and ancient trace fossils show that they can be divided into a number of assemblages specific to different sedimentary environments (Seilacher, 1964; 1967). The assemblage of the Um Sahm shales is a classic example of the near-shore tidal *Cruziana: Scolithos* suite (p. 18). Accordingly the fine-grained sub-facies of the Um Sahm may be attributed to the Z zone environment shorewards of the sand shoals, rather than to a deeper off-shore low-energy X zone.

The Um Sahm Formation of Jordan is therefore a marine shelf deposit. This account demonstrates that the X, Y, Z zone concept can be usefully applied to non-carbonate facies.

To conclude this discussion of marine shelf deposits the following points should be noted:

(1) It is important to distinguish tectonic shelves and basins from sedimentary shelf and basin environments. They are not synonymous.

(2) A tectonic shelf is a stable part of the earth's crust which can be a site of erosion or deposition. Sedimentation may take place in various environments ranging from continental, through shorelines to open marine.

(3) A shelf sedimentary environment occurs on sub-marine tectonic shelves.

(4) A tectonic basin is a large, essentially synclinal, part of the earth's crust (though it may actually be convex to the sky when allowance is made for the curvature of the earth, Dalmus, 1958).

(5) If subsidence occurred faster than sedimentation then deep water deposits may infill the basin (as in the Permian Delaware basin of west Texas, see p. 215).

(6) A slowly subsiding tectonic basin may be infilled by shallow marine shelf, shoreline, or conventional deposits.

(7) The open marine shelf environment is therefore only one of several that may occur on a tectonic shelf. It may also prevail over a gently subsiding tectonic basin.

The deposits of a marine shelf are of three main types. Below wave base calcilutites and shales form under low-energy conditions. Where wave base impinges on the gently sloping sea floor high-energy clean-washed sands (carbonate or clastic) form shoals and barriers. Up slope of these a second low-energy environment is found in lagoons and tidal flats. Here deposition may be argillaceous or, if there is no land-derived sediment, pelletal limestones, dolomites, and evaporites may form.

This concept of three main shelf environments is equally applicable to gently subsiding regions such as the Williston basin and to tectonically more complex areas like the Sirte basin. It is relevant to both carbonate environments, as in these two cases, and to terrigenous shelf deposits such as the Um Sahm Formation of Jordan.

In areas like the Williston basin the evaporites were clearly shallow water (?diagenetic supratidal) deposits which migrated basinwards through time.

Evaporites occur in the centres of many carbonate sedimentary basins such as the Silurian Michigan basin, and the Permian basins of west Texas, and the North Sea. Their situation in such depocentres has led to their being interpreted as originating on the floor of deep restricted marine basins (e.g. Borchert and Muir, 1964, p. 43). It is interesting to speculate, however, whether some of these are actually shallow Z zone deposits of shelf environments (see Shearman and Fuller, 1969; and Kirkland and Evans, 1973).

## ECONOMIC ASPECTS

Shelf deposits are of great economic significance. Many of the world's major oil fields occur in shelf carbonates of Iraq, Iran, Libya, and the Williston basin (see Dunnington, 1958; Falcon, 1958; Colley, 1963; and Carlson and Anderson, 1965).

These accounts show that oil accumulates largely in the Y zone shoal carbonates and associated reefs (see p. 234). These often retain their primary porosity (some Libyan reservoirs occur in completely unconsolidated carbonate sand). Alternatively, they may have been cemented but secondary porosity can form by dolomitization and leaching.

The nearshore low-energy Z zone facies generally has less favourable reservoir characteristics. Porosity may sometimes occur in microcrystalline dolomites and pellet limestones. Fracturing can be important in upgrading the reservoir potential of this facies. The caprocks of shelf carbonate reservoirs are generally sabkha evaporites, though micritic chalks can also fill this role.

X zone micrites are seldom prospective. They may have porosity, as in chalk, but generally lack permeability unless they are extensively fractured. The Ekofisk and associated fields of the North Sea produce from fractured Cretaceous chalk reservoirs (Byrd, 1975; Scholle, 1977; D'Heur, 1984).

The regionally persistent stratigraphy of shelf carbonates is of great significance since it allows the formation of laterally extensive reservoirs. For example, shoal limestones of the Upper Jurassic Arab-D in Saudi Arabia are host to the Ghawar oil field which is over 30 km long. The gentle but regionally persistent dips of carbonate shelf deposits allow the local accumulation of hydrocarbon reservoirs from vast catchment areas.

It can be seen, therefore, that carbonate shelf deposits are of some interest in the search for oil and gas. The X-Y-Z concept is a useful exploration tool which aids the prediction of optimum reservoir characteristics in such rocks. Wilson (1975), and Reeckmann and Friedman (1982) give further details of the application of facies analysis to the exploration for petroleum in carbonate reservoirs.

Carbonate shelf deposits are economically interesting for other reasons since they can contain evaporites and phosphates. The origin of phosphates is a complex and controversial problem and the environmental parameters which control their formation are not well understood (e.g. Bromley, 1967). Valuable phosphate deposits occur in Upper Cretaceous shelf carbonates of the Middle East. These are found in a belt stretching from Syria (the Soukhne Formation) through the Wadi Sirhan of Saudi Arabia to Jordan, Palestine, Egypt, and Morocco (Tooms *et al.*, 1971). Though it is clear from their facies relationships that these phosphates are marine deposits their precise environment of formation seems to be variable. In Egypt, Said attributes them to regressive shoreline conditions (1962, pp. 251, 268) while Youssef attributes them to deposition in syn-sedimentary basins on the deeper parts of the sea bed (Youssef, 1958). In Jordan on the other hand the phosphatic facies is restricted to palaeohighs (Bender, 1968, p. 189). There is an interesting problem here (Notholt, 1980; Batutin, 1982).

## SUB-SURFACE DIAGNOSIS OF SHELF DEPOSITS

The techniques for diagnosing shelf deposits vary greatly according to whether they are carbonate or terrigenous (Heckel, 1972).

As pointed out in Chapter 1, carbonate environments are largely diagnosed by petrography; terrigenous environments largely by sequences of sedimentary structures and grain size.

Thus sub-surface diagnosis of carbonate rocks is largely made from the microscopic examination of well cuttings, aided by cores, when they are available.

The position within the X-Y-Z spectrum of a particular lithology may be determined by its texture and grain-type as shown in Figs 8.1 and 8.2. Moving from basin centre to basin margin the texture of the carbonate will change in the following sequence: micrite, wackestone, packstone, grainstone (or boundstone in reef shorelines), packstone, micrite. Correspondingly there are variations in grain type from lime mud (with pelagic fossils), skeletal grains of benthonic fossils and/or ooliths, followed by peloidal grains in the Z zone lagoons and tidal flats. These are of course gross over-simplifications and environmental diagnosis of carbonate rocks should not be attempted without practical studies of modern carbonate sediments and microscopic petrography of ancient ones.

Terrigenous shelf deposits by contrast present a different set of problems. In many ways they are indistinguishable from barrier bar sands, particularly where a marine sand overlies an unconformity and is itself overlain by transgressive marine shales (see for example Swie-Djin, 1976).

Studies of modern shelf seas with strong tidal currents, such as the North Sea

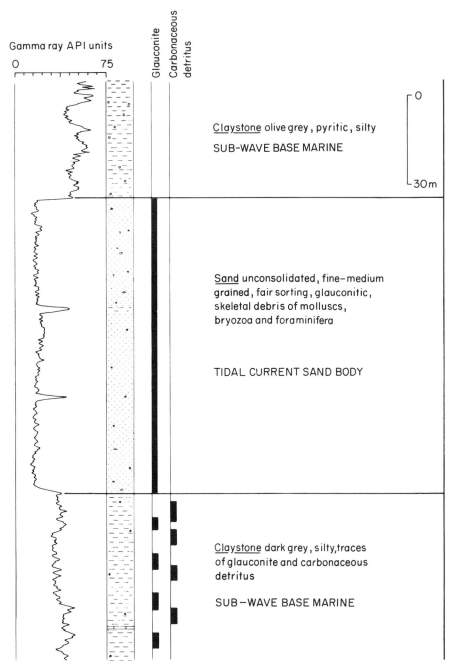

*Figure 8.15* Borehole penetrating a tidal current sand body. Note the glauconitic, shelly lithology, the abrupt scoured base and the uniform clean gamma log. From Selley (1976) Fig. 7, by courtesy of the *American Association of Petroleum Geologists.*

and South East Asia, have drawn attention to sand ridges which migrate across current scoured bed-rock.

These can be ten of metres thick, hundreds of metres in width and thousands of metres in length, being aligned parallel to the axis of tidal ebb and flow (Stride, 1970; Reineck, 1971). They are composed of shelly glauconitic sands. Sparker surveys show that they have complex internal cross-stratification reflecting the complex current systems from which they were deposited.

Tidal current sand bodies such as these can sometimes be determined in the sub-surface from seismic data. These 'sand build-ups', as they are sometimes vulgarly termed, show planar bases, complex internal reflections, and corrugated upper surfaces. Fig. 8.15 shows a borehole log through a sand body of inferred tidal current origin.

## REFERENCES

The description of the Williston basin was based on:

Andrichuk, J. M., 1958. Mississippian Madison Stratigraphy and Sedimentation in Wyoming and Southern Montana. In: *Habitat of Oil*. (Ed. L. C. Weeks) Amer. Assoc. Petrol. Geol. pp. 225–67.

Carlson, C. G., and Anderson, S. B., 1965. Sedimentary and tectonic history of North Dakota, part of Williston Basin. *Bull. Amer. Assoc. Petrol. Geol.*, **49**, 1833–46.

Darling, G. B., and Wood, P. W. J., 1958. Habitat of oil in Canadian portion of Williston Basin. In: *Habitat of Oil*. (Ed. L. G. Weeks) Amer. Assoc. Petrol. Geol. 129–48.

Edie, R. W., 1958. Mississippian sedimentation and oil fields in south-eastern Saskatchewan. *Bull. Amer. Assoc. Petrol. Geol.*, **42**, 94–126.

Illing, L. V., Wood, G. V., and Fuller, J. G. C. M., 1967. Reservoir rocks and stratigraphic traps in non-reef carbonates. *Proc. Seventh World Petrol. Cong.*, Vol. 2, 487–99.

Irwin, M. L., 1965. General theory of epeiric clear water sedimentation. *Bull. Amer. Assoc. Petrol. Geol.*, **49**, 445–59.

Hansen, A. R., 1966. Reef trends of Mississippian Ratcliffe zone, northeast Montana and northwest North Dakota. *Bull. Amer. Assoc. Petrol. Geol.*, **50**, 2260–8.

Harris, S. H., Land, C. B., and McKeever, J. H., 1966. Relation of Mission Canyon stratigraphy to oil production in north-central North Dakota. *Bull. Amer. Assoc. Petrol. Geol.*, **50**, 2269–76.

Proctor, R. M., and Macauley, G., 1968. Mississippian of Western Canada and Williston Basin. *Bull. Amer. Assoc. Petrol. Geol.*, **52**, 1956–68.

Wendell Smith, G., Summer, G. E., Wallington, D., and Lee, J. L., 1958. Mississippian oil reservoirs in Williston Basin. In: *Habitat of Oil*. (Ed. L. G. Weeks) Amer. Assoc. Petrol. Geol. 149–77.


Other references cited in this chapter were:

Ball, M. M., 1967. Carbonate sand bodies of Florida and the Bahamas. *Jl. Sediment. Petrol.*, **37**, 556–91.

Baturin, G. N., 1982. *Phosphorites on the sea floor; origin, composition and distribution.* Elsevier, Amsterdam, p. 344.

Bebout, D. G., and Pendexter, C., 1975. Secondary carbonate porosity as related to Early Tertiary depositional facies, Zelten field, Libya. *Bull. Amer. Assoc. Petrol. Geol.*, **59**, 665–93.

Bender, F., 1968. *Der Geologie von Jordanien.* Beitrager Reg. Geol. Erde. Bd. 7, Borntrager, Berlin.

——, 1975. Geology of the Arabian Peninsula-Jordan. *U.S. G.S. Prof. Pap.*, 560–i, p. 136.

Bennacef, A., Beuf, S., Biju-Duval, B., de Charpal, O., Gariel, O., and Rognon, P., 1971. Example of Cratonic sedimentation: Lower Paleozoic of Algerian Sahara. *Bull. Amer. Assoc. Petrol. Geol.*, **55**, 2225–45.

Borchert, H., and Muir, R. O., 1964. *Salt Deposits, the origin, metamorphism and deformation of evaporites.* Van Nostrand, London, p. 338.

Bouma, A. H., Berryhill, H. L., Knebel, H. J., and R. L. Brenner, 1982. Continental Shelf. In: *Sandstone Depositional Environments.* (Eds P. A. Scholle, and D. Spearing). Amer. Assoc. Petrol. Geol. Tulsa. 281–328.

Bromley, R. G., 1967. Marine phosphorites as depth indicators. In: *Depth indicators in marine sedimentary environments.* (Ed. A. Hallam) *Marine Geol.*, Sp. Issue, **5**, No. 5.6, 503–10.

Byrd, W. D., 1975. Geology of the Ekofisk field, offshore Norway. In: *Petroleum and the Continental Shelf of Northwest Europe.* (Ed. A. W. Woodland) Applied Science Publishers, London. pp. 439–46.

Colley, B. B., 1964. Libya: Petroleum Geology and development. *Sixth World Petrol. Cong. Proc.* Section 1, pp. 1–10.

Conley, C. D., 1971. Stratigraphy and lithofacies of Lower Paleocene rocks, Sirte Basin, Libya. In: *Symposium on the Geology of Libya.* (Ed. C. Gray) University of Libya, Tripoli. pp. 127–40.

Dalmus, K. F., 1958. Mechanics of basin evolution and its relation to the habitat of oil in the basin. In: *Habitat of Oil.* (Ed. L. G. Weeks) Amer. Assoc. Petrol. Geol. pp. 883–931.

Dunnington, H. V., 1958. Generation, migration, accumulation, and dissipation of oil in Northern Iraq. In: *Habitat of Oil.* (Ed. L. G. Weeks) Amer. Assoc. Petrol. Geol. pp. 1194–252.

Emery, K. O., 1968. Relict sediments on continental shelves of the world. *Bull. Amer. Assoc. Petrol. Geol.*, **52**, 445–64.

Enos, P., 1982. Shelf. In: *Carbonate Depositional Environments.* (Eds P. A., Scholle, D. B. Bebout and C. H. Moore) Amer. Assoc. Petrol. Geol. Mem. **33**, 267–96.

Evans, G., Schmidt, V., Bush, P., and Nelson, H., 1969. Stratigraphy and geologic history of the sabkha. Abu Dhabi, Persian Gulf. *Sedimentology*, **12**, 145–59.

Falcon, N. L., 1958. Position of oil fields of southwest Iran with respect to relevant sedimentary basins. In: *Habitat of Oil* (Ed. L. G. Weeks) Amer. Assoc. Petrol. Geol. pp. 1279–93.

Fischer, A. G., 1964. The Lofer cyclothem of the Alpine Triassic. *Kansas Geol. Surv. Bull.* **169**, 107–49.

Hakansson, E., Bromley, R., and Perch-Nielsen, K., 1974. Maastrichtian chalk of northwest Europe, a pelagic shelf sediment. In: *Pelagic sediments: on land and sea.* (Ed. K. J. Hsu and H. C. Jenkyns) Blackwell Scientific Publications, Oxford, pp. 221–33.

Heckel, P. H., 1972. Recognition of Ancient shallow marine environments. In: Recognition of Ancient sedimentary environments. *Soc. Econ. Pal. Min.*, Sp. Pub. No. 16. (Eds J. K. Rigby and W. K. Hamblin) pp. 226–86.

Helal, A. H., 1965. Stratigraphy of outcropping Palaeozoic rocks around the northern edge of the Arabian Shield (within Saudi Arabia). *Z. Deutsch. Geol. Ges.* Band 117, pp. 506–43.

D'Heur, M. 1984. Porosity and Hydrocarbon distribution in the North Sea chalk reservoirs. _Mar & Pet. Geol._ **1**. 211–39.

Holser, W. T., 1966. Diagenetic polyhalite in recent salt from Baja, California. _Am. Mineralogist_, **51**, 99–109.

Illing, L. V., 1954. Bahaman calcareous sands. _Bull. Amer. Assoc. Petrol. Geol._, **38**, 1–95.

Jefferies, R., 1963. The stratigraphy of the _Actinocamax plenus_ subzone (Turonian) in the Anglo-Paris Basin. _Proc. Geol. Ass._, **74**, 1–34.

Kennedy, W. H., 1980. Aspects of Chalk sedimentation in the southern Norwegian offshore. In: _The Sedimentation of North Sea reservoir Rocks._ Norwegian Pet. Soc. Symp., Oslo.

Kinsman, D. J., 1969. Modes of formation, sedimentary association and diagnostic features of shallow-water and supratidal evaporites. _Bull. Amer. Assoc. Petrol. Geol._, **53**, 830–40.

Kirkland, D. W., and Evans, R., 1973. _Marine evaporites: origins, diagenesis and geochemistry._ Benchmark papers in geology. Dowden, Hutchinson & Ross. Inc., Strondsburg, Penn, p. 426.

Klitzsch, E., 1966. Comments on the geology of the central parts of southern Libya and northern Chad. In: _South Central Libya and Northern Chad._ (Ed. J. J. Williams) Petrol. Explor. Soc. Libya. 1–19.

Lees, A., 1973. Platform carbonate deposits. _Bull. Centre Rech. Pau._, **7**, 177–92.

McManus, D. A., 1975. Modern versus relict sediment on the continental shelf. _Bull. Geol. Soc. Amer._, **86**, 1154–60.

Meischner, K. D., 1965. Allodapische Kalke, turbidite in Riff-Nahen Sedimentations-Becken. In: _Turbidites._ (Eds A. Bouma and A. Brouwer) Elsevier, Amsterdam. pp. 156–191.

Newell, N. D., Purdy, E. G., and Imbrie, J., 1960. Bahaman oolitic sand. _Jour. Geol._, **68**, 481–97.

Notholt, A. J. G., 1980. Economic phosphatic sediments: mode of occurrence and stratigraphic distribution. _Jl. Geol. Soc. Lond._, **137**, 793–805.

Purdy, E. G., 1963. Recent calcium carbonate facies of the Great Bahama Bank. _Jour. Geol._, **71**, 334–55 and 477–97.

Purser, B. H. (Ed.), 1973. _The Persian Gulf (Holocene carbonate sedimentation and diagenesis in a shallow epicontinental sea)._ Springer-Verlag, Berlin, p. 471.

Reeckmann, A. and Friedman, G. M., 1982. _Exploration for Carbonate Petroleum Reservoirs._ J. Wiley & Sons, Chichester, p. 213.

Reineck, H. E., 1971. Marine sandkörper, rezent und fossil (Marine sandbodies, recent and fossil). _Geol. Runsch._, **60**, 302–21.

Said, R., 1962. _The Geology of Egypt._ Elsevier, Amsterdam, 377 p.

Scholle, P. A., 1977. Chalk diagenesis and its relation to petroleum exploration: oil from the Chalk, a modern miracle. _Amer. Assoc. Petrol. Geol. Bull._, **61**, 982–1009.

Seilacher, A., 1964. Biogenic sedimentary structures. In: _Approaches to Paleoecology._ (Eds J. Imbrie and N. D. Newell) Wiley, N.Y. pp. 296–316.

——, 1967. Bathymetry of trace fossils. In: Depth indicators in marine sedimentary environments. (Ed. A. Hallam) _Marine Geol._, Sp. Issue, **5**, No. 5/6, 413–28.

Selley, R. C., 1970. Ichnology of Palaeozoic Sandstones in the Southern Desert of Jordan: a study of trace fossils in their sedimentologic context. In: _Trace Fossils._ (Eds J. C. Harper, and T. P. Crimes) Lpool. Geol. Soc. 477–88.

——, 1972. Diagnosis of marine and non-marine environments from the Cambro-ordovician sandstones of Jordan. _Jl. Geol. Soc. Lond._, **128**, 109–17.

——, 1976. Sub-surface environmental analysis of North Sea sediments. _Bull. Amer. Assoc. Petrol. Geol._, **60**, 184–95.

Sharma, G. D., 1972. Graded sedimentation on Bering shelf. In: *24th Internat. Geol. Cong. Montreal.* Section 8, 262–71.

Shearman, D. J., 1963. Recent anhydrite, gypsum, dolomite and halite from coastal flats of the Arabian shore of the Persian Gulf. *Proc. Geol. Lond.*, No. 1067, 63–5.

——, 1966. Origin of marine evaporites by diagenesis. *Bull. Inst. Min. Met.*, **76**, section B., 82–6.

——, and Fuller, J. G. C. M., 1969. Phenomena associated with calcitization of anhydrite rocks, Winnepegosis Formation, Middle Devonian of Saskatchewan, Canada. *Proc. Geol. Soc. Lond.*, No. 1658, 235–9.

Shinn, E. A., 1984. Tidal Flat. In: *Carbonate Depositional Environments.* (Eds P. A. Scholle, D. G. Bebout and C. H. Moore). *Amer. Assoc. Petrol. Geol. Mem.*, **33**, 171–210.

Stanley, D. J., and Swift, D. H. P. (Eds), 1976. *Marine sediment transport and environmental management.* Wiley Interscience, New York. p. 602.

Stride, A. H., 1970. Shape and size trends for sandwaves in a depositional zone of the North Sea. *Geol. Mag.*, **107**, 469–78.

——, 1982. *Offshore Tidal Sands.* Chapman and Hall, London, p. 260.

Swie-Djin, N., 1976. Marine transgressions as a factor in the formation of sandwave complexes. *Geol. en Mijnb.*, **55**, 18–40.

Swift, D. J. P., Duane, D. B., and Orrin, H. P., 1973. *Shelf sediment transport: process and pattern.* Wiley, Chichester, p. 670.

Tebbutt, G. E., Conley, C. D. and Boyd, D. W., 1965. Lithogenesis of a distinctive carbonate rock fabric. *Univ. Wyoming Contrib. Geol.*, **4**, 1–13.

Tooms, J. S. Summerhayes, C. P., and McMaster, R. L., 1971. Marine geological studies on the northwest African margin: Rabat-Dakar. In: *The geology of the east Atlantic continental margin*, **4**, Africa. Inst. Sci. Rept. 70/16, pp. 11–25.

Wilson, J. L., 1975. *Carbonate Facies in Geologic History.* Springer-Verlag, Berlin, p. 471.

Youssef, M. I., 1958. Association of phosphates with synclines and its bearing on prospecting for phosphates in Sinai. *Egypt. J. Geol.*, **2**, 75–87.

# 9  Reefs

## INTRODUCTION

The term 'reef' was originally applied to rocky prominences on the sea floor on which ships could be wrecked. Coral reefs are a particular form of this navigational hazard found today in tropical waters.

Geologists have applied the term reef to lenses made of the calcareous skeletons of sedentary organisms.

In many cases, however, it is not possible to demonstrate either that an ancient 'reef' was a topographic high on the sea floor or, if it was, that it was wave resistant.

Cummings (1932) classified calcareous skeletal deposits into:

*Bioherm* 'A reef, bank, or mound; for reeflike, moundlike, or lens-like or otherwise circumscribed structures of strictly organic origin, embedded in rocks of different lithology' (*ibid.*, p. 333).

*Biostrome* 'Purely bedded structures such as shell beds, crinoid beds, coral beds, et cetera, consisting of and built mainly by sedentary organisms, and not swelling into moundlike or lens-like forms . . . which means a layer or bed' (*ibid.*, p. 334).

Subsequently the term 'reef' has been applied very loosely to lenses of carbonate, often even on the basis of seismic data prior to drilling. Dunham (1970), Braithwaite (1973), and Heckel (1974) have reviewed the problems of reef identification and classification. It is extremely difficult to place a lens of carbonate rock into one of the rigidly defined classes that some nomenclatural schemes propose. This is not only because there is a limited amount of rock available for study, in the form of cuttings and cores, but also because diagenesis may have destroyed the critical evidence of fauna and texture on which a definition of 'reef' may hinge.

For the purposes of this book the following definitions will be used:

*Carbonate build-up:* A lens of carbonate rock (a descriptive term). Carbonate build-ups may be divided into two genetic types.

*Reef:* A carbonate build-up of skeletal organisms which at the time of formation was a wave resistant topographic feature which rose above the general level of the sea floor.

*Bank:* A carbonate build-up which was a syn-depositional topographic high of non-wave resistant material, e.g. an oolite shoal, a coquina bank, or a mound of crinoid debris.

This chapter describes ancient carbonate build-ups which have been interpreted as reefs. First, though, the features of Recent reefs will be sumarized. Coral reefs are one of the best documented of all present-day environments. The following brief synthesis is based on the references listed at the end of this chapter.

## Summary of Recent reefs

The majority of reefs growing at the present time occur in shallow tropical seas. The factors which restrict reef growth are variable but in general they form best in less than 25 fathoms of water where the salinity is between 27–40 per thousand and the temperature seldom drops below about 20°C (Shepard, 1963, p. 351). The inevitable exception to the rule is where coral reefs form in cold water 35 fathoms deep off the Norwegian coast (Teichert, 1958). Reefs of calcareous algae occur in non-marine waters (Fouch and Dean, 1982).

The biota of Recent reefs is exceedingly diverse (Jones and Endean, 1973). The resistant reef framework is composed of corals, calcareous algae, hydrocorallines, and bryozoa. Corals are generally actually subordinate to the other groups (Yonge, 1973). Other organisms associated with reefs include calcareous sponges, foraminifera, echinoids, lamellibranchs, gastropods, and sabellariid (carbonate-secreting) worms.

In cross-section a reef complex can be broadly sub-divided into four geomorphological units (Fig. 9.1).

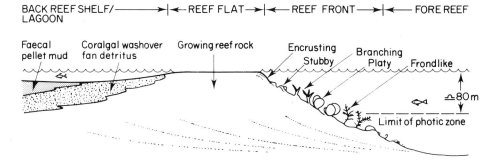

*Figure 9.1* Cartoon through a reef to show the distribution of sediment types, and the relationship between ecology and environment.

A reef generally shelters a shallow water lagoon from the open sea. The floor of the lagoon is covered by carbonate mud in its deeper parts and sands in shallower turbulent regions. Lagoonal sediments are composed of faecal pellets, foraminiferal sands, coralgal sands of comminuted corals and calcareous algae, together with other skeletal sands and finely divided carbonate mud. Scattered through the lagoon may be irregular patch reefs. Grain size increases across the lagoon towards the reef and locally becomes conglomeratic due to organic debris broken off the reef and carried into the lagoon by storms.

The reef itself is, by definition, composed of a resistant framework of calcareous organic skeletons. The top of the reef is flat since the creatures cannot survive prolonged sub-aerial exposure. In addition the upper surface is constantly scoured and planed off by wave action and dissected by seaward trending surge channels. The tops of these sometimes become overgrown to form submarine tunnels. The reef framework itself is often highly porous, figures of up to 80% porosity have been cited (Emery, 1956).

The seaward edge of the reef, the reef front, is a submarine cliff with a talus slope at its base. This slope is made up of organic debris broken off from the reef front; grain size decreases down slope into deeper water. Immediately at the foot of the reef front boulders of reef rock may be present, grading into sand and then mud in deeper water. The talus slope supports a fauna similar to that of the reef but the quieter conditions allow more delicate branching corals and calcareous algae to form. The reef talus has poorly developed seaward dipping bedding. Slumps and turbidites have been described from ancient reef flanks (p. 221 and p. 249).

It is important to put reefs in perspective as far as their size is concerned. Reefs may perhaps be best thought of as machines for making carbonate

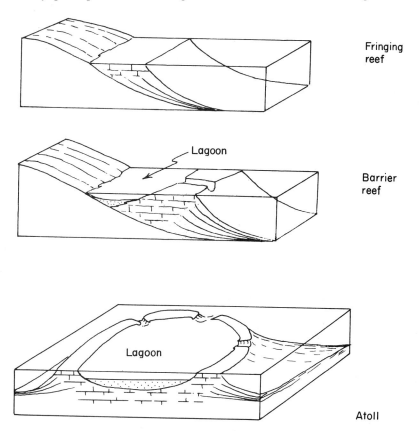

Fringing reef

Lagoon

Barrier reef

Lagoon

Atoll

*Figure 9.2* Illustrative of the three main types of present day reefs.

sediment. The biogenic carbonate material which grows in the reef environment is constantly being attacked by the sea. Thus relatively small areas of actual reef seem to be preserved within vast formations of coralgal sands and muds derived from them. This point is illustrated by the Great Barrier Reef of Australia. Though this modern reef tract extends for some 2000 km along the coast, few individual reefs are more than 20 km in length (Maxwell, 1968; Hopley, 1982).

On the basis of their geometry recent reefs are classified into three main kinds. Inevitably there are transitions between the three types.

*Fringing reefs* are linear in plan and stretch parallel to coasts with no intervening lagoons (Fig. 9.2). Fringing reefs can form where low rainfall means that little freshwater and mud is brought into the sea to inhibit the growth of reef colonial organisms. Good examples occur along the desert shores of the Gulf of Akaba on the Red Sea (Friedman, 1968).

*Barrier reefs* are linear too, but a lagoon separates them from the land (Fig. 9.2). This may be narrow or, in the case of the Great Barrier Reef of Australia, an open sea hundreds of kilometres wide (Maxwell, 1968).

*Atolls* are sub-circular reefs enclosing a lagoon from the open sea (Fig. 9.2). This kind of reef is abundant in the Pacific; Bikini Atoll is a typical example (Emery, Tracey, and Ladd, 1954). The classic theory for atoll formation was put forward by Darwin in 1842. This proposed that barrier reefs originally formed around volcanic islands. As the island sank under its own weight the barrier would build upwards to keep pace with the relative rise in sea level. Ultimately after all trace of the volcanic island had vanished beneath the sea a circular coral reef, or atoll, would remain.

Studies of Recent reefs and the concepts derived therefrom can be usefully applied to the study of ancient reefs (Laporte, 1974; Toomey, 1981; James, 1983). Before describing case histories of fossil reefs it is necessary however to make three cautionary remarks. First, the reef-building organisms of the past were not always of the same groups as those of today. The roles of different groups vary in time and place. For example, calcareous algae have in some situations merely acted as binding agents holding the skeletons of the reef framework organisms together. In other situations and at other times algae formed the dominant framework of reefs.

The second point to note is that Recent reefs have suffered subaerial exposure during the Pleistocene Period. Thus their present day topography is in part a drowned one. In particular it should be noted that most modern reef fronts are old sea cliffs.

The third point concerns the relationship between Recent and ancient reef geometries. True fringing reefs appear to be rare in the geological record. Linear reef complexes are common but either are of the barrier type separating open marine deposits from lagoonal facies or lie on structural highs between basinal troughs.

Fossil atolls are rare. Sub-circular reef complexes have been described as atolls but are not based on volcanic piles. Examples include the Horseshoe Atoll and Scurry-Snyder reefs (Pennsylvanian) of west Texas (Ellison, 1955; Leverson, 1967, pp. 73 and 327) and the El Abra reef of Mexico. This was originally interpreted as a barrier, but offshore studies have now revealed an oval trend (Guzman, 1967).

The second major type of ancient reef, in addition to barriers, is essentially a circular reef core rimmed by reef slope talus. The core may provide shelter for a lagoon on its leeward side. This type of structure is termed a patch reef, or if markedly conical in vertical profile, a pinnacle reef.

Case histories of ancient barrier and atoll reefs will now be described.

## PERMIAN REEFS OF WEST TEXAS: DESCRIPTION

One of the best documented ancient carbonate shelf margins occurs in the Permian rocks of west Texas (Figs 9.3 and 9.4). These crop out around the rim of the Delaware basin and have been traced eastwards at sub-surface around the edge of the Midland Platform where they contain important oil reservoirs. Because of their economic interest these reefs have been intensively studied at outcrop to aid sub-surface interpretation of facies and trends. The following account is a summary based on the references cited at the end of this chapter.

A series of sedimentary facies can be recognized which are broadly concentric around the basin margin and which prograded basinwards one above the other. These are summarized in Table 9.1.

Three facies are found on the northwestern shelf; the Bernal, Chalk Bluff and Carlsbad facies. These occur in laterally extensive interbedded formations. The Bernal facies, which is best developed in the north, consists of red laminated siltstones with desiccation cracks and interbedded cross-laminated and laminated red sandstones.

The overall fine grain size of the Bernal faces indicates a low energy environment. The desiccation cracks and red colour point to subaerial exposure and continental conditions. It seems most probable, therefore, that this facies was deposited in a complex of vast shallow lagoons or playa lakes.

Traced south across the shelf the continental Bernal facies interfingers with the Chalk Bluff facies. This consists of interlaminated gypsum and dolomite wackestone. The transitional position of these evaporites with the marine carbonates to the south suggests that the Chalk Bluff facies was deposited in an intertidal salt marsh environment.

The third and most basinward of the three shelf facies is the Carlsbad. This can be divided into two sub-facies. The first consists of dolomitic wackestones and packstones with abundant peloidal grains. It also contains algal stromatolites. The latter are indicative of a shallow marine environment, the ill-sorted textures suggest a poorly winnowed environment, and the abundance of pelloidal grains is characteristic of modern coastal lagoons and restricted

Table 9.1. *Summary of facies and environments of the Permian rocks of west Texas,* *based on the schemes of Meissner (1972), Wilson (1975), and other workers. The* *Delaware basin turbidites and the Bernal red beds were deposited during regressive* *phases, and the other facies during marine transgressions.*

| FACIES NAME | DESCRIPTION | ENVIRONMENT |
|---|---|---|
| | | *North* |
| Bernal | Red siltstone | Continental |
| Chalk Bluff | Laminated gypsum and dolomite wackestone | Sabkha evaporite |
| Carlsbad | Dolomite wackestone, algal stromatolites, peloidal dolomites | Restricted shelf |
| | Cross-bedded skeletal dolomitic grainstones | Exposed shelf margin |
| Capitan | Skeletal wackestones and boundstones | Reef or shelf margin |
| | Major slope-bedded wackestones and breccias | Slope margin |
| Delaware | Radiolarian black shales | Starved basin |
| | Fine grained sandstones | Turbidites |
| | | *South* |

marine shelf environments. The second of the two Carlsbad sub-facies consists of cross-bedded dolomitic skeletal grainstones. The cross-bedding and well sorted texture of this facies suggest deposition on an open marine shelf.

Between the three platform facies just described, and the basin, lies the Capitan facies in the shelf margin position. This is divided into two subfacies. Along the crest of the scarp is a massive carbonate unit. This is composed of boundstone (*in situ*) reef rock, and of skeletal grainstones and wackestones. The fossils of this facies are of both sessile benthos and encrusting colonial organisms. These include sponges, bryozoa, calcareous algae, crinoids and brachiopods. True corals are rare. This fauna is ecologically zoned with respect to the shelf margin (Fig. 9.5).

This massive sub-facies of the Capitan prograded basinward over the second sub-facies. This consists of skeletal wackestones, breccias and boulder beds which show basinward-directed depositional dips of up to 30° (Fig. 9.4). The key to the interpretation of the west Texas Permian lies in the massive sub-facies of the Capitan. The inclined-bedded sub-facies clearly formed on a submarine slope between the shelf edge and the basin floor. The debate on the origin of the massive Capitan limestone hinges on to what extent it may have been a true barrier reef. This will be discussed in the next section.

The toe-sets of the Capitan slope sub-facies interfinger basinward with the Delaware facies. This is nearly four thousand feet thick and covers about one

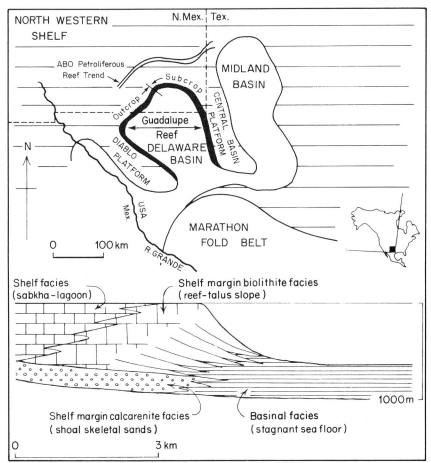

*Figure 9.3* Upper: Permian tectonic framework of west Texas. Capitan shelf margin shown in black. Modified from King, 1934, p. 704, *Bulletin of the Geological Society of America*. Lower: diagrammatic cross-section illustrating facies distribution in the Permian rocks of west Texas. Modified from Newell, N. D., *et al.*, *The Permian Reef Complex of the Guadalupe Mountains, Texas and New Mexico* Fig. 5.D. W. H. Freeman and Company. Copyright 1953. The shelf facies includes the Bernal Chalk Bluff and Carlsbad facies. The shelf margin consists of the Capitan Limestone. The Basinal facies consists of the sandstones and shales of the Delaware Group.

hundred thousand square miles of basin floor. Two interbedded sub-facies are recognizable. One consists of black carbonaceous radioactive laminated claystones and siltstones. These are occasionally interbedded with fine-grained limestones containing pelagic foraminifera, ammonoids, siliceous sponge spicules and radiolaria.

Traced to the slope margin limestones become more common and coarser grained, and contain fragments of derived skeletal material from the shelf and shelf margin.

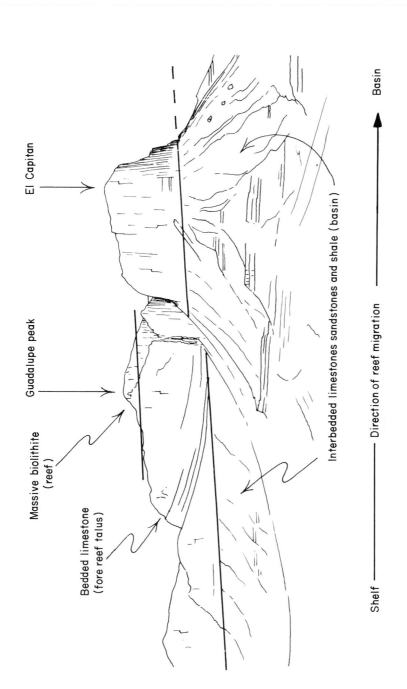

El Capitan

Guadalupe peak

Massive biolithite
(reef)

Bedded limestone
(fore reef talus)

Interbedded limestones sandstones and shale (basin)

Shelf ————————— Direction of reef migration ————————— Basin

*Figure 9.4* Photograph and sketch of the Guadalupe Mountains, West Texas, showing facies relationships. Photo by courtesy of J. C. Harms, P. N. McDaniel, and L. C. Pray.

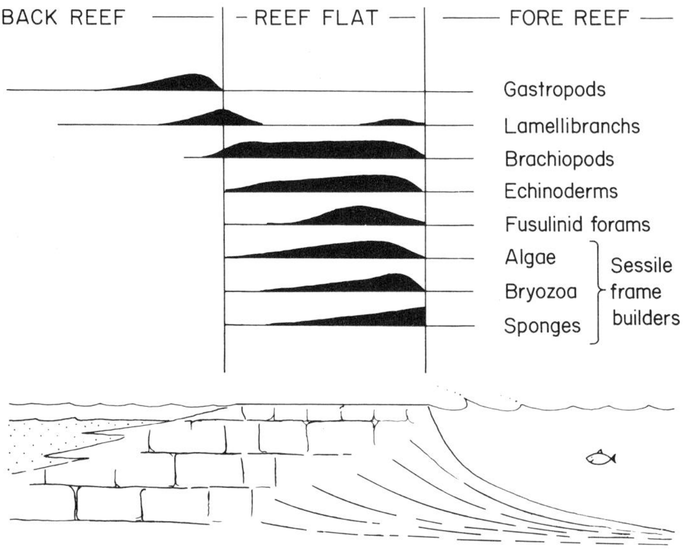

*Figure 9.5* Inferred life distribution of west Texas Permian reef organisms. From Newell, N. D., *et al.*, *The Permian Reef Complex of the Guadalupe Mountains, Texas and New Mexico* Fig. 79 W. H. Freeman and Company. Copyright 1953.

The second sub-facies of the Delaware facies is of terrigenous clastics. This is composed of fine-grained, graded and laminated sandstones whose petrography suggests that they were derived from granitic source areas which lay to the north of the basin, on the other side of the shelf margin. Particular attention has been paid to these sandstones because they are hydrocarbon reservoir rocks. Using closely spaced wells it has been shown that they occur in elongated pods (channels?) which are aligned perpendicularly to the basin margin.

The Delaware facies was clearly laid down in relatively deep water, certainly below the limits of wave and tidal action. The black shales were most probably deposited when the basin was starved of sediment and the circulation of its water was restricted. The graded sandstone beds have been interpreted as turbidites (see Chapter 10) brought in from beyond the Northwestern Shelf, while the skeletal limestones were brought in, perhaps by the same process, from the carbonate shelf margin.

Considered overall, therefore, the Permian rocks of west Texas provide a classic example of a carbonate shelf, slope, basin progradational complex, which was ultimately infilled by evaporites.

## PERMIAN REEFS OF WEST TEXAS: DISCUSSION

There is general agreement that the carbonate Permian rocks of west Texas were deposited in shelf, slope and basinal environments. After the earlier work of King (1934) a major study was carried out by Newell *et al.* (1953). These workers interpreted the massive facies of the Capitan Formation as a huge barrier reef which extended for several hundred miles around the shelf edge. Subsequent more detailed studies have questioned this interpretation.

The basic facts of geometry are undisputed: that a major environment of organic carbonate formation occurred along the shelf margin, that this prograded basinward over a talus slope of its own detritus, and that the maximum depositional relief from the top of the shelf to the basin floor was in the order of 700 m.

The point of discussion is whether the shelf margin was a barrier reef. The answer to this question lies partly in detailed palaeontological and petrographic studies, and partly on one's definition of the term 'reef'. There is general agreement that the massive Capitan facies contains typical reefal fauna, both framework building types, and binding and encrusting forms to hold the framework together. There is argument as to whether this boundstone lithology is present in sufficient amounts to have formed a living barrier (Kendall, 1969, p. 2507). The total volume is difficult to quantify, partly because of the difficulties of access in this mountainous terrain, and partly because extensive recrystallization makes it hard to discern the original depositional fabric.

Achauer (1969, p. 2317) also accepts the presence of *in situ* algally bound biolithite, but doubts that this lacked the ecological potential ever to have formed a wave resistant topographic feature (*ibid.*, p. 2321). This raises another interesting point. Many workers have noted that the Capitan facies exhibits many signs of subaerial exposure. These include vadose pisolites (Dunham, 1969), desiccation polygons termed 'tepee' structures, intraformational channels and conglomerates, and diagenetic cements similar to those found in modern subaerially exposed limestones. Thus Dunham (1970) distinguishes between 'ecological' reefs which are organically bound, and 'stratigraphic reefs' whose wave resistant topography is due to inorganic cementation. He considers the Capitan to be a stratigraphic reef. This idea of the shelf being subaerially exposed has been pursued by Meissner (1972). Though these facies are essentially regressive, prograding as they do into the Delaware basin, they also show cyclic sedimentation. Meissner (*ibid.*) has suggested that during transgressions carbonate sedimentation on the shelf was accompanied by anaerobic starved conditions in the basin. These transgressions alternated with regressive drops in sea level. At these times the shelf margin carbonates were subaerially exposed and cemented (forming a stratigraphic reef). Continental conditions prevailed on the shelf, with terrigenous sediment transported basinward through channels cut in the exposed shelf margin to be deposited as turbidites on the basin floor.

This mechanism suggests, therefore, that the massive Capitan limestone is not a true ecological barrier reef, though it may have acted intermittently as a stratigraphic one. The consensus of opinion now seems to be that the Capitan massive limestone formed in slightly deeper water below the shelf crest, where delicate framework-building organisms acted as sediment baffles, trapping carbonate mud. Intermittent exposure caused the cementation of these mud-mounds, which then formed a stratigraphic reef at the commencement of the next transgression; the lithified surface of which would be an ideal substrate for encrusting and other benthonic organisms. Indeed detailed observation shows that some foreset bedding persists through the massive Capitan limestone and that the crest of the shelf margin may actually have been where the Carlsbad cross-bedded skeletal grainstones now lie (Wilson, 1975, p. 228) In other words, the Northwestern Shelf was separated from the Delaware basin sometimes by marine shoal sands, and sometimes by cemented carbonate barrier islands, but seldom if ever by a true ecological barrier reef.

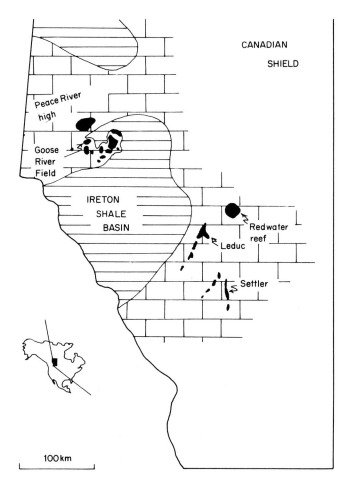

*Figure 9.6* Devonian facies distribution in southern Alberta. Canada. Modified from Jenik and Lerbekmo, 1968, Fig. 3. From the *Bulletin of the American Association of Petroleum Geologists*, by courtesy of the American Association of Petroleum Geologists.

As can be seen from Fig. 9.3 the various Permian facies just described migrated through time from shelf to basin. This can be clearly seen from the way in which the massive biolithite overlies the dipping bedded limestones. More dramatic still is the way that the evaporitic and red-bed facies migrate from the back of the shelf out to the centre of the Delaware basin where they constitute the thick Ochoan Series (Hills, 1968, Figs. 4–7). The lower part of the Ochoan Series (the Castile and Salado formations) is composed of several thousand feet of laminated anhydrite, polyhalite, gypsum, halite, and shales. These are often varved, contorted, and with nodular fabrics. Individual shale units can be traced over hundreds of square miles. This evaporite facies is overlain by fine red sandstones and siltstones of the Rustler and Dewey Lake formations.

## LEDUC (DEVONIAN) REEFS OF CANADA: DESCRIPTION

The Leduc reef complex provides many good examples of patch reefs, the second common type of ancient reefs. These developed on the edges of the Ireton shale basin (Devonian) of Alberta (Fig. 9.6). Patch reefs occur on the break in slope along the edge of the shelf and on the shelf itself. Since oil was discovered in great quantities in 1947, large amounts of data have been gathered from intensive drilling. The Leduc carbonate lenses overlie the laterally continuous Cooking Lake Formation. This is composed of limestones, sometimes argillaceous, and thin shale interbeds. The limestones are sometimes bioclastic calcarenites and sometimes biostromes with thin bioherms of small lateral extent. The top of the Cooking Lake Formation is a planar surface on which the Leduc Formation carbonate build-ups are based. The crest of individual lenses sometimes lie 300 m above the top of the Cooking Lake Formation (e.g. the Acheson Field area). The majority of the Leduc structures lie along the northeast-southwest slope break on the eastern edge of the Ireton Shale basin. This trend contains the Acheson, Leduc, Bonnie Glen, and Rimby oil fields. The Redwater field occurs in an isolated build-up on the shelf to the east. Another archipelago of reefs of slightly older age occurs on the edge of the Peace River palaeohigh on the northwest slope of the Ireton basin. Petrographically the majority of the Leduc carbonates are medium- to fine-grained vuggy crystalline dolomites and contain few detectable fossils. These are several exceptions to this rule which retain their primary features, one of these will be described in detail shortly. The Leduc carbonate lenses are separated from one another laterally by the Duvernay Formation. This is a thinly-bedded sequence of dark bituminous pyritic limestones and shales. Limestones, largely argillaceous, are best developed in the lower part of the formation and contain a fauna of crinoids, ostracods, and brachiopods. Calcareous shales are dominant towards the top of the Duvernay Formation with a fauna of sponge spicules, brachiopods, ostracods, conodonts, and plant spores.

The Goose River carbonate build-up is one of several that lie on the northwestern shelf of the Ireton basin. It is sub-circular in plan; being about 12 km in diameter from southwest to northeast, and 8 km across from southeast to northwest. It is approximately 50 m high. A total of twenty-two rock types and eight environmental facies have been recognized in this body of rock. For the purpose of this account these have been greatly condensed. Essentially the Goose River build-up consists of a marginal facies, sub-circular in plan, which encloses a central facies, Fischburch (1968), and Jenik and Lerbekmo (1968).

The marginal facies consists of biolithites, calcarenites, and calcirudites. The biolithite is made up largely of *in situ* stromatoporoid colonies with massive and short stubby branching morphologies. Calcarenites occur as lenses interbedded with the biolithite and in radial channels cross-cutting it. They are composed of fragmented stromatoporoids, bryozoa, brachiopods, and echinoderms (both crinoids and echinoids). Intraformational conglomerates are found on both the outer and inner edges of the marginal facies. These consist of angular and rounded pebbles set in an argillaceous matrix. The pebbles are micritic with scattered entire and fragmented algae, stromatoporoids, bryozoa, crinoids, echinoids, brachiopods, and foraminifera.

The three major lithologies of the marginal facies just described enclose a fine-grained carbonate facies in the centre of the build-up. This consists of laminated micrites, micritic pellet and skeletal sands, and pisolitic and oncolitic algal beds. The biota is different from that of the marginal facies. Macrofossils are largely absent. Microfossils are more common; including foraminifera, ostracods, and calcispheres. (The latter are possibly the reproductive cysts of dasycladacean algae.) Stromatoporoids are present, as in the marginal facies, but have delicate branching morphologies as shown by the form *Amphipora* (Fig. 9.7).

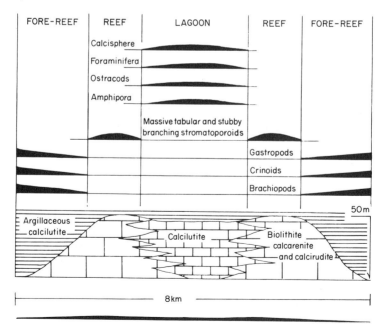

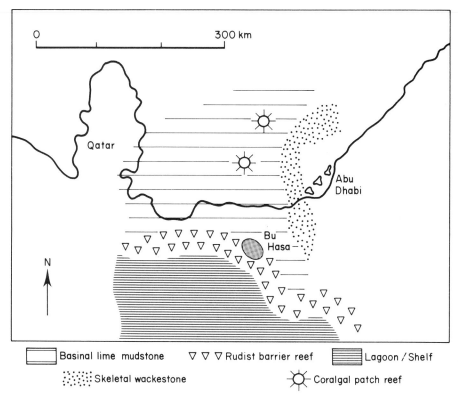

Basinal lime mudstone ▽ ▽ ▽ Rudist barrier reef ▤ Lagoon / Shelf

░░ Skeletal wackestone ☼ Coralgal patch reef

*Figure 9.8* Map to show the regional setting of the Cretaceous Bu Hasa reef oil field, Abu Dhabi. Based on Harris *et al.* (1968), Twombley and Scott (1975) and Hassan *et al.* (1975).

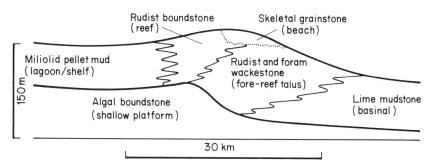

*Figure 9.9* Cross-section to show the facies and environments (in brackets) of the Bu Hasa reef oil field, Abu Dhabi. Based on Harris *et al.* (1968), Twombley and Scott (1975) and Hassan *et al.* (1975).

*Figure 9.7 (opposite page)* Cross-section of major sedimentary facies and faunal distribution in the Goose River Reef (Upper Devonian) of Alberta, Canada. Condensed greatly from Jenik and Lerbekmo, 1968, Figs 8, 9a, and 10a; from the *Bulletin of the American Association of Petroleum Geologists*, by courtesy of the American Association of Petroleum Geologists.

## LEDUC (DEVONIAN) REEFS OF CANADA: INTERPRETATION

The geometry of the Leduc Formation shows that it consists of a series of carbonate build-ups, by definition. Where intensive diagenesis has obliterated primary fabrics it is not possible to prove whether these were banks or reefs. It is only in unaltered examples, such as the Goose River build-up, that their true origin can be determined.

The Leduc and Duvernay Formations are overlain by the Ireton Shale Formation which is locally draped as it thins over Leduc palaeohighs. This formation, over 200 m thick, consists of laminated pyritic argillaceous dolomites and dolomitic shales with thin shell beds.

The Goose River reef will now be interpreted. This is a good example of the Devonian reefs of Alberta though it slightly predates the Leduc reefs previously described. It is atypical of many of these since it is not a crystalline dolomite; primary depositional fabric and fauna have thus been preserved.

The marginal stromatoporoid biolithite facies indicates that the Goose River lens is a true reef. From this initial conclusion the origins of the other facies fall easily into place. The calcirudites and calcarenites associated with the marginal biolithite are probably due to fragmentation of the reef by wave action. The resultant detritus was washed over the reef crest into the lagoon, or allowed to settle on an outer talus slope.

The fine-grained sediments in the centre of the build-up indicate a lower-energy environment in which microfossils and more delicate macrofossils could thrive. The lithology, fauna, and medial distribution of this facies indicate that it formed a lagoon enclosed by the sub-circular reef. Access to the open sea was probably provided by the channels cross-cutting the marginal facies which are infilled by skeletal calcarenites.

A similar, though less symmetrical distribution of facies has also been described from the Redwater build-up on the southeastern shelf of the Ireton basin. In this example a northwest-southeast trending arcuate reef sheltered lower-energy deposits to the south-west. There was no totally enclosed lagoon.

By analogy these two reefs suggest that the other, now recrystallized, build-ups also originated as patch reefs in which arcuate or circular atoll-like reef cores sheltered low-energy environments from the open sea.

## THE BU HASA REEF OIL FIELD, ABU DHABI: DESCRIPTION

The Bu Hasa oil field was discovered in onshore Abu Dhabi. It produces from the Aptian Shuaiba Formation at the top of the Thamama Group. The Bu Hasa field occupies a local culmination on a Cretaceous carbonate shelf margin that developed between the Arabian shield to the south and the Gulf trough to the north (Fig. 9.8).

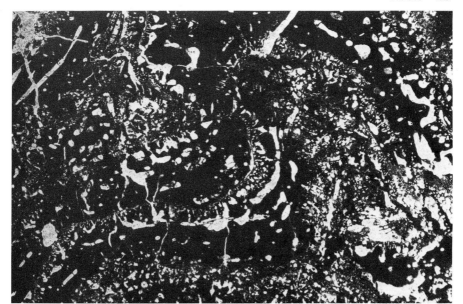

*Figure 9.10a* Photomicrograph of the algal boundstone lithology on which the Bu Hasa reef developed.

*Figure 9.10b* Photomicrograph of lime mudstone deposited in the basin to the north of the Bu Hasa reef.

Detailed petrographic studies defined a large number of carbonate microfacies. A simplified cross-section showing their distribution across the field is shown in Fig. 9.9. The Bu Hasa carbonate buildup growth was initiated on a laterally

*Figure 9.10c*  Photomicrograph of Rudist reef rock from the Bu Hasa field.

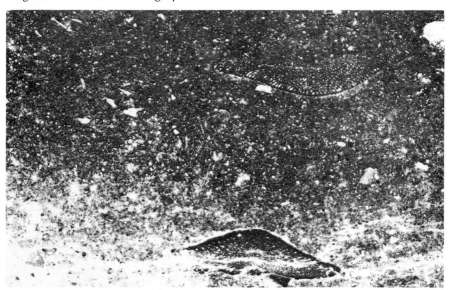

*Figure 9.10d*  Photomicrograph of forereef skeletal wackestone with Orbitoline forams and Rudist debris.

extensive biostrome formed by the encrusting calcareous alga *Bacinella*. This unit is overlain on the shelf by *Lithocodium* boundstone (Fig. 9.10a). This passes basinward into argillaceous, bituminous lime mudstones with occasional pelagic foraminifera (Fig. 9.10b).

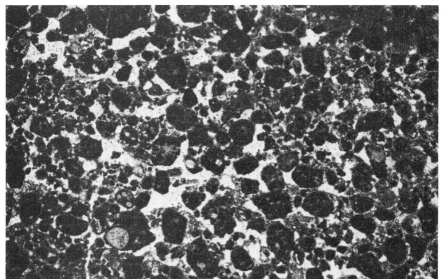

*Figure 9.10e* Photomicrograph of faecal pellet packstone from the shelf behind the Bu Hasa barrier reef.

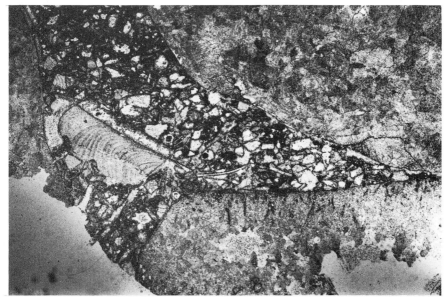

*Figure 9.10f* Photomicrograph of Rudist grainstone from the seaward rim of the Bu Hasa reef.

The basinward edge of the Lithocodium unit is overlain by a rudist boundstone (Fig. 9.10c). This is about 60 m thick and 6 km wide and can be traced laterally along the shelf margin for some 200 km. (The rudists were a group of lamellibranchs that assumed a coralline habit, with one valve resembling the calyx and the other a protective lid. They were extremely successful reef builders in the Cretaceous Period, and provide petroleum reservoirs in Libya, Mexico and USA as well as in the Gulf). The Bu Hasa rudist boundstones pass basinward into skeletal wackestones, with fragments of rudists and large benthonic Orbitoline forams (Fig. 9.10d). These wackestones pass basinward into lime mudstones. The rudist boundstone passes south towards the Arabian shield into faecal pellet muds, with miliolid foraminifera. (Fig. 9.10e). Locally the basinward crest of the rudist boundstone is replaced by a detrital rudist grainstone (Fig. 9.10f).

The top of the Shuaiba formation is a regional unconformity. It is overlain by the organic-rich shales of the Albian Nahr Umr Formation. This is the presumed source for the oil in the Bu Hasa and adjacent fields.

## THE BU HASA REEF OIL FIELD, ABU DHABI: INTERPRETATION

The *Bacinella* and *Lithocodium* boundstones were clearly deposited in shallow water, within the photic zone required for algal development. The lateral transition of the *Lithocodium* boundstone into the lime mudstones suggest the initiation of topographic relief on the sea bed. The lime mudstones, with their pelagic forams must have been deposited in quieter deeper water below the limit photic zone. As with so many reefs, the seaward limit of the shelf favours the establishment and ebullient reproduction of carbonate-secreting organisms. Thus the seaward rim of the *Lithocodium* boundstone appears to have provided ideal conditions for the development of a rudist barrier reef. The Orbitolina wackestones on the seaward side of the reef presumably represent the fore-reef talus that graded out into the basinal lime mudstones. The Miliolid-bearing faecal pellet muds behind the reef are analogous to the lagoonal deposits of recent carbonate shelf environments.

The skeletal grainstones along the seaward rim of the reef front are perhaps beach deposits. The pre-Albian unconformity marks an extensive event of subaerial exposure during which solution porosity developed across the underlying Shuaiba Formation. The skeletal grainstones along the reef rim were perhaps formed during the regression. The present structural relief of the Bu Hasa dome measures some 35 by 20 km. It is not clear to what extent this relief is structural, depositional or erosional. The distribution of porosity and permeability in the field locally cross cuts the facies boundaries, and is in part due to solution associated with the post-Shuaiba unconformity.

This case history thus demonstrates how petrography aids the environmental interpretation of carbonate environments. It also shows, alas, that facies analysis does not necessarily aid the prediction of porosity and permeability in carbonate petroleum reservoirs.

## GENERAL DISCUSSION OF REEFS

In this section the following three topics are discussed: factors controlling the geometries and facies distribution of reefs, the association of reefs with evaporites and euxinic shales, and lastly, the diagenesis of reef rocks.

# Factors controlling reef geometries and facies

The shape of a reef and the distribution of associated facies are controlled by the interplay of sea-level changes, tectonic setting, biota, and oceanography.

It is widely held that ancient reefs grew in shallow water (absolute depth unspecified). This conclusion is based on analogy with Recent reefs. Furthermore, the sedimentary facies associated with ancient reefs generally suggest shallow shelf environments.

If it is true that ancient reefs grew in shallow water their growth must have been closely controlled by fluctuations of sea level. Reef organisms cannot build up above the sea since they are killed by prolonged sub-aerial exposure. Similarly they cannot grow at great depths because algae, being plants, only thrive within the photic zone. Many living ceolenterates are similarly restricted because they have symbiotic relationships with zooxanthellae (a type of algae) within their living tissues.

When the sea level remains static the reef will prograde seaward over its own talus slope, as in the case of the Permian reefs of west Texas. If sea level rises slowly the reef will build essentially upwards with no lateral facies migration, or it will transgress landward over the back reef lagoonal facies. Examples of this occur in the Devonian reefs of the Canning basin, Australia (Playford and Lowry, 1966, p. 8). A rapid rise in sea level will kill a reef due to great depth. A slow drop in sea level will cause the reef to migrate seaward and downward. However, such sequences are relatively rare since, as the shoreline retreats, old reefs tend to be destroyed by sub-aerial erosion. A rapid drop in sea level will kill a reef instantaneously due to prolonged exposure. It can be seen therefore that fluctuating sea level exerts an important effect on the geometry of a reef and its associated facies.

The second main controlling factor of reef geometry is tectonic setting, since this is also important in bringing the sea bed in crucial juxtaposition to sea level. Basically reefs are typical of tectonic shelves where sedimentation is shallow, marine, and free from land-derived clastics. Within this broad realm four main sub-types can be recognized. First reefs very commonly form at the edge of a shelf where it passes into a deeper basin. Along such trends barrier reefs may form (e.g. The Permian reefs of west Texas) or discontinuous lines of patch reefs such as the Leduc (Devonian) complex of Canada. Sometimes the edge of the shelf may be a fault with reefs built along the crest of the fault scarp. The Cracoe and Craven reefs of north England may be a case in point (Bond, 1950).

Sometimes a shelf is too deep for reef-formation but local syn-sedimentary movement along anticlinal crests bring the sea bed within a depth sufficiently shallow for reefs to develop. The Palaeocene Intisar reefs of Libya are one example of this (Terry and Williams, 1969, p. 45), the Clitheroe reefs in the Lower Carboniferous Bowland trough of England are another (Parkinson, 1944).

Similarly volcanic eruptions on the sea floor can build piles of lava up into shallower depths where their crests may be colonized by reef organisms. The Devonian Hamar Lagdad reef of the Tafilalet basin, Morocco, is of this type (Massa and others, 1965). As already mentioned, ancient atoll-capped volcanoes are rare. This is possibly because such features are essentially oceanic and unlikely to be preserved in areas at present accessible to the land-based geologist.

The fourth arrangement of reefs on shelves is where patch reefs are distributed at random over a wide area. Such archipelagoes, or buckshot patterns as they are sometimes colourfully called, occur in the Niagaran (Silurian) beds on the edge of the Canadian Shield (Lowenstam, 1950).

The third controlling factor of reef geometry and facies to consider is the biota. The essential control of water depth has been noted already. One of the curious features of reefs is that though they range in age from PreCambrian to the present day, and show similar geometries and sub-facies, their fauna varies through time. PreCambrian algal biostromes have been classified into several types according to their geometry and internal anatomy. Opinions are divided as to whether these differences are of stratigraphic or ecological control (Logan, Rezak, and Ginsburg, 1964). Lower Cambrian rocks in many parts of the world contain reefs built by curious beasts of uncertain affinities termed Archaeocyathines. Reefs are widespread in Palaeozoic rocks throughout the world. The framework of these was built principally by stromatoporoids, rugose corals, hydrocorallines, and bryozoa. Rugose and tabulate corals largely became extinct at the end of the Palaeozoic Era, while the alcyonaria and hexacorals suddenly developed at that point in time and occupied the same ecological niche. In Lower Cretaceous times a particular group of lamellibranchs, the Rudistids, became important reef builders. In these forms one valve became shaped like the calyx of a simple coral while the other formed the lid. Rudistid reefs of Lower Cretaceous age occur in the Edwards Limestone of Texas and the Golden Lane of Mexico (Griffith, Pitcher, and Wesley Rice, 1969), also in the Alps, Libya, and Iran (Henson, 1950). The Richtofenid group of brachiopods followed a similar pseudo-coralline reefal habit in the Upper Palaeozoic. Through the vast span of geological time there seem to have been few aquatic invertebrate groups which have not at some time or another developed colonial reef-forming species. At almost all times, however, the calcareous algae have been important reef organisms (Harlan Johnson, 1961). Generally their role has been to bind together reef frameworks composed of various other organisms. Sometimes, though, they form the framework of reefs largely unaided by other groups. Algal reefs generally tend to be thinner and laterally more continuous than those of other creatures.

Hadding (1950) and Manten (1971) have described how the geometry of Silurian reefs of Gotland was controlled by their fauna. Three types are present. Lensoid reef knolls are built of spheroidal stromatoporoids, *Favosites* and *Heliolites*. Biostromes are made of tabular stromatoporoids. Pinnacle reefs, with well-developed flank breccias, have cores of delicate branching columnar stromatoporoids.

It can be seen therefore that, since a reef is composed almost entirely of fossils, palaeontology and palaeoecology are vital to an understanding of their depositional environment. Furthermore, since the biota of a reef controls its diagenesis and hence porosity development, an understanding of reef biota is of some economic significance.

## The reef: evaporite: euxinic shale association

The Permian rocks of west Texas show a pattern of stagnant basinal sediments rimmed by reefs and overlain by evaporites. This association is a common one in ancient sediments. Weeks (1961) lists nineteen other examples all of which, like the Permian complex, are petroliferous. There has been considerable speculation on the origin and development of such basins. The conventional explanation is that the reefs grew around the margin of a restricted sea where upper waters of normal salinity encouraged reef growth. Dense brines in deeper water inhibited benthonic activity allowing foul-smelling muds to form and ultimately precipitated evaporite minerals on the deeper parts of the basin floor. As the basin filled up the evaporite facies transgressed shorewards killing the reefs and trapping within them hydrocarbons which had migrated from the basinal muds (Hunt, 1967, p. 237). Critical to this concept is the demonstration that marginal reef growth was synchronous with basinal evaporite precipitation (for examples, see Henson, 1950, Fig. 11 and Heybroek, 1965, pp. 28–9).

This is not always possible to prove, however, and an alternative mechanism is sometimes possible. It has already been described how evaporites can form at the present time due to diagenesis of inter-tidal and supratidal sabkhas (p. 188). These deposits are not only petrographically similar to those of 'basinal' evaporites but many of the textures are identical too, such as organic laminations, contortions, and anhydrite nodules. Thus Shearman and Fuller (1969) have discussed the possibility that inter-reef evaporites of the Devonian Winnepegosis Formation of Canada may be diagenetic sabkha deposits rather than basin floor precipitates. If this is correct then the evaporites formed after the reefs were killed by a drop in sea level.

Similarly, Permian evaporites of the Delaware basin advanced through time from the shelf to the basin (Hills, 1968, Figs 4–7). This is what one would expect from a prograding sabkha. One would expect the reverse effect if the evaporites formed on the basin floor. This prompts the question, have tectonic basins (gently subsiding areas infilled with sediment) been confused with sedimentary basins (basins which at the time of sedimentation were topographic sub-marine depressions) (see p. 203).

Clearly the origin of basinal evaporites and their time relationships with reefs deserves careful evaluation (see also Kirkland and Evans, 1973).

## Diagenesis of reefs

Strictly speaking the diagenesis of reefs is post-depositional and therefore falls outside the study of depositional environments which is the theme of this book.

However, reef rock diagenesis is critical to an understanding of their economic significance. It deserves some attention, therefore. The diagenesis of carbonates in general, and of reefs in particular, is a large and complex topic which has been discussed at length in the literature (see for example Bathurst, 1975; Chilingar, Mannon, and Rieke, 1972; Langres, Robertson, and Chilingar, 1972; Reeckmann and Friedman, 1982). The following account is therefore necessarily brief.

Reefs show three interesting and rather unusual properties which are of great significance to their post-depositional history. First, they are formed with a very high primary porosity. Second, they are lithified at their time of formation, compaction is slight and initially primary porosity is preserved. Three, reefs are formed of chemically unstable minerals (dominantly aragonite and calcite). These can undergo diverse chemical changes due to reactions with circulating pore fluids. Basically two types of diagenetic changes can be recognized in reefs: mineralogical and textural. These are obviously inter-related. Initially, aragonitic skeletal material alters to its stable polymorph calcite. Theoretically this should result in an increase in volume and a concomitant decrease in porosity and permeability (Hoskin, 1966, p. 1072). This process takes place at different rates in different areas of the reef complex due to the different stabilities of the various carbonate particles (Friedman, 1964). Simultaneously, or later than the aragonite: calcite change, the carbonates may be enriched with magnesium from the sea water and converted to dolomite. Theoretically this should in contrast result in an overall contraction of the total rock volume of as much as 13% causing intercrystalline porosity between dolomite crystals (Chilingar and Terry, 1964). Silicification, often involving the selective replacement of fossils, is another diagenetic process common in reefs (Newell *et al.*, 1953, p. 173).

These chemical changes are associated with textural modification of the reef fabric. First, primary porosity may be reduced shortly after deposition due to the infiltration of finely divided carbonate mud, produced by the fragmentation of calcareous algae. A reduction of porosity may also be achieved through infilling of the framework by sparry calcite accompanied by recrystallization of the carbonate skeletons and the development of calcite overgrowths. These changes all result in a decrease in porosity and, together with dolomitization, often destroy all signs of the original organic fabric. Subsequently secondary porosity may occur due to solution along fractures accompanied by the formation of biomoldic and vuggy porosity. Many of these processes are reversible.

It can be seen therefore that the diagenesis of a reef is complex, being a function of its original faunal and lithological composition and of the chemistry of the fluids which subsequently circulated within it. The main points to be noted, however, are that recrystallization may completely destroy the primary organic fabric of a reef. Secondly, though ancient reefs are often highly porous, the type and distribution of the porosity may bear no relation to that which existed at the time of their formation. This is demonstrated by the Bu Hasa oil field case history. These points are critical to an understanding of the economic geology of reefs.

## ECONOMIC GEOLOGY OF ANCIENT REEFS

Ancient reef deposits are of great economic significance: because of their porosity they often make good hydrocarbon reservoirs when suitably sealed, and because of their porosity and chemical instability they are liable to replacement by metallic ore minerals.

The association of reefs rimming shale basins and capped by evaporites has been already noted (p. 233). Considerable attention has been given to the occurrence of oil in such situations. It is generally agreed that the evaporites play a critical role both because their parent brines inhibited oxidation of organic matter, allowing petroleum to form and because the evaporites themselves make good caprocks (Week, 1961; Sloss, 1955). The significance of carbonates as oil source rocks in their own right has been frequently discussed (e.g. Hunt, 1967). However, it is generally considered that petroliferous reefs are unlikely to be their own source rocks due to intensive oxidation of organic matter during reef-formation. It is interesting to note, however, that in Tertiary reefs of Iran, oil trapped within fossils predates cementation and is different in composition from that which fills the bulk of the reservoir (Henson, 1950).

Apart from the reef itself, oil and gas may accumulate in overlying non-reef porous strata draped over the reef crest. For additional data on reefal oil reservoirs see Harbaugh (1967, pp. 382–95) and Reeckman and Friedman (1982).

Though oil and gas accumulations in reefs are hard to find and exploit the effort can be rewarding.

Apart from their ability to trap large quantities of oil and gas, reefs are also the host to metalliferous deposits. These are of a particular variety termed Mississippi Valley type after one of the prime examples (see Evans, Campbell, and Krouse, 1968; and Brown, 1968; Hutchison, 1983 p. 47–62). These are telethermal replacement sulphide ore deposits of sphalerite and galena, with subordinate fluorspar, barite, dolomite, and, of course, calcite. Typically these occur replacing reefs and other carbonate build-ups in tectonically stable areas far distant from faults and igneous rocks which could have been sources for the metals. The role of evaporites in providing briny solutions to transport the metals into reefs has been discussed by Davidson (1965). Amstutz in particular has championed a sedimentary origin for the source of many of these ores (Amstutz, 1967, p.431; Amstutz *et al.*, 1964).

As the mud around a reef compacts, residual metal-rich solutions are expelled and may escape into adjacent porous reefs where they replace the host rock.

As with oil and gas the distribution of minerals within a reef complex varies from place to place. Lower Carboniferous reefs of Ireland have mineralized cores (Derry, Clark, and Gillatt, 1965). The Devonian reefs of the Canning basin, Australia, are mineralized in the fore-reef facies (Johnstone and other, 1967), while in Alpine Triassic reefs mineralization sometimes occurs in the lagoonal back-reef facies.

It can be seen therefore that ancient reefs are of considerable economic significance both as hydrocarbon reservoirs and as hosts for mineralization. Sub-surface facies analysis has a role to play in the location and exploitation of these deposits.

## SUB-SURFACE DIAGNOSIS OF REEFS

Considerable attention has been given to methods of locating oil fields in ancient reefs. This is not always easy. The main problems of identifying ancient reefs are twofold. It is hard to locate reefs since they may be scattered at random over shelves buried beneath horizontal strata often with no surface structural expression. It is hard to identify an oil pool as a reef since diagenesis has often obliterated the original organic fabric. These two problems will now be considered.

Since normal methods of surface geological field work are often inapplicable, reef hunting is largely based on geophysical techniques. This is often largely a matter of luck. Searching for pinnacle reefs only a few kilometres in diameter in several hundred square kilometres of shelf can be like looking for a needle in a haystack. A case in point was the discovery of the Intisar reefs in the Sirte basin of Libya. The second seismic line to be shot in a 1880 sq km$^2$ concession just happened to transect the crest of a reef. The discovery well was tested for an initial production of 43 000 bd. The seismic programme was initiated to search for structures. The reef was a shock (Terry and Williams, 1969).

Reefs, being composed of limestone, or occasionally dolomite, are generally denser than the shales and evaporites with which they are commonly overlain. Reefs may thus be identified as a positive Bouguer anomaly on gravity surveys (Ferris, 1972). This method has now largely been superseded by the seismic method.

Reefs can generally be identified from seismic data. Where there is sufficient velocity contrast between the reef and its cover then the top may show up as a reflecting horizon. This is generally the case when a reef is capped by evaporites. With compacted shales however the velocity contrast may be too small to generate a signal. Sometimes a reflector is present over the flank of a reef but dies out over the crest. This implies that the crestal limestone is acoustically slower, and therefore more porous, than on the flanks. Reefs may be seismically invisible if there is no velocity contrast with the overlying sediments. Domed reflectors above the level of a reef may indicate where it lies (Boulware, 1967).

Reefs themselves tend to have a uniform velocity, and thus generally do not produce any internal reflectors. An exception to this rule may be a 'bright spot'

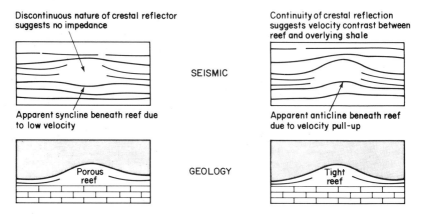

*Figure 9.11* Cartoons to show some of the tricks involved in the seismic interpretation of reefs. For explanation see text.

indicative of a fluid contact (generally gas on liquid). Examples have been noted in some of the Upper Permian Zechstein reefs in the North Sea (Jenyon and Taylor, 1983). Normally however reflectors in the flank of a reef die out towards the core.

The reflecting horizon beneath a reef may also be interesting. A tight reef may cause a velocity 'pullup' suggesting to the unwary that the reef overlies an anticline. By contrast the reverse situation of an apparent syncline beneath a reef is good news. It signifies a velocity 'pulldown' suggesting a low-velocity interval and hence porosity. Figure 9.11 illustrates some of these examples. For further details see Bubb and Hatledid (1977).

With good quality seismic data, and where the velocity contrast is sufficient, the geometry of reefs may be mappable. Figures 9.12 and 9.13 illustrate examples of pinnacle and barrier reef respectively.

If geophysical data is absent or of poor quality reefs may be located from well data. Criteria to look for are anomalous thickening of limestones between closely spaced wells, anomalous dips due to reef talus bedding, and faunal changes suggesting the proximity of an organic build-up (e.g. Andrichuk, 1958, p. 90). The reef core, if drilled through, may be unrecognizable due to recrystallization. Where two wells are drilled, one of which passes through an open marine facies and the equivalent interval of the other is lagoonal, it may be profitable to search for a barrier reef in the intervening ground.

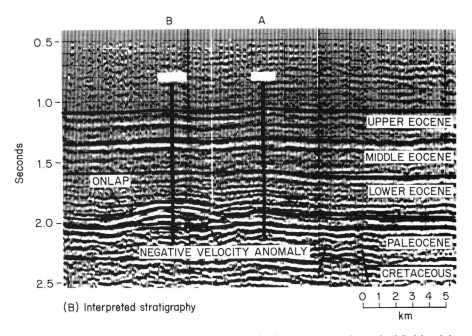

*Figure 9.12* Seismic line through one of the Idris/Intisar pinnacle reef oil fields of the Sirte basin, Libya. Note negative velocity anomaly beneath the reef. From Bubb and Hatledid 1977, Fig. 5, by courtesy of the American Association of Petroleum Geologists.

The case history of the Bu Hasa oil field (pp. 226–230) demonstrates how petrography holds the key to the subsurface facies analysis of reefs. This example also demonstrates how diagenesis can alter the original distribution of porosity and permeability.

Thus, even when a reef has been identified it may be hard to predict the distribution of optimum reservoir properties within it. In some cases production comes from the reef core itself (e.g. the Leduc and Intisar reefs of Canada and Libya). In other situations the reef core may be barren and production restricted to the fore-reef facies as in Tertiary reefs of Iran (Henson, 1950). Back-reef lagoonal facies are generally poor reservoirs. In pelletal limestones porosity may be high but permeability low (Stout, 1964, p. 334). This may be increased by fracturing. Oil production comes from lagoonal sands of the Permian Texas reef complex where they interfinger up dip with evaporites and dolomites (Leverson, 1967, p. 292).

As mentioned earlier, gamma and S.P. log motifs are of little use in the diagnosis of carbonate environments. An upward-decreasing A.P.I. curve may be present on the gamma curve where a fore-reef talus progrades transitionally over basinal shales. The overlying reef will show the typical pure carbonate log profile of a steady low A.P.I. curve (Fig. 9.14).

The dipmeter is particularly useful, however, in establishing the geometry

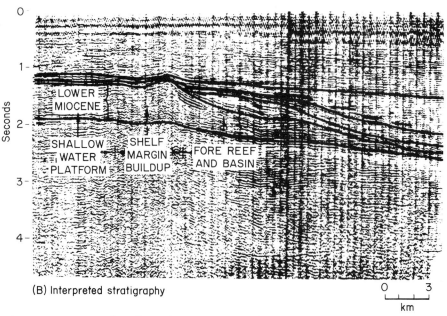

(B) Interpreted stratigraphy

*Figure 9.13* Seismic line through a Miocene barrier reef complex in the Gulf of Papua. Note how the offlapping nature of some of the forereef talus suggests emergence and therefore the possible existence of solution porosity on the reef crest. From Bubb and Hatledid, 1977, Fig. 7. By courtesy of the American Association of Petroleum Geologists.

of a reef once it has been found. The reef core often shows a random 'bag o' nails' motif due to the fractured and recrystallized fabric. Depositional dip is often clearly seen in the fore-reef talus slope, with a characteristic upward-increasing blue pattern. The reef core lies in the opposite direction to the progradational dip (Fig. 9.14). Where two adjacent wells have both inadvertently penetrated the flank of a pinnacle reef, its crest may be found by drawing two lines reciprocal to the fore-reef dip direction (Jageler and Matuszak, 1972, p. 128).

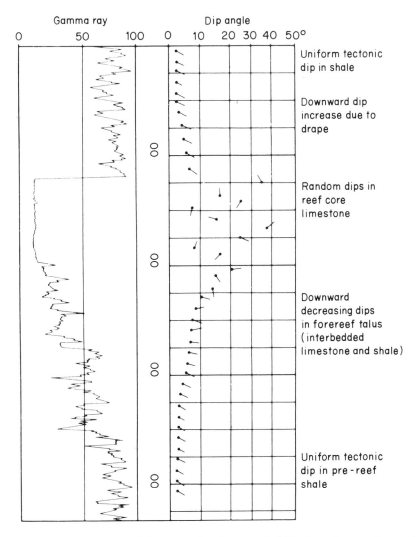

*Figure 9.14* Gamma ray and dipmeter log for a well drilled on the southeast side of a 'reef'.

# REFERENCES

This is an exceedingly brief bibliography of works on Recent reefs:

Emery, K. O., Tracey, J. I., and Ladd, H. S., 1954. Geology of Bikini and nearby atolls. Pt. I. Geology. *U. S. Geol. Surv. Prof. Pap.*, 260–A, p. 265.

Fairbridge, R. W., 1950. Recent and Pleistocene coral reefs of Australia. *J. Geol.*, **58**, 336–401.

Friedman, G. M., 1968. Geology and geochemistry of reefs, carbonate sediments and waters, Gulf of Akaba (Elat) Red Sea. *J. Sediment. Petrol.*, **38**, 895–919.

Ginsburg, R. N., Lloyd, R. M., Stockman, K. W., and McCallum, J. S., 1963. Shallow water carbonate sediments. In: *The Sea*, vol. III. (Ed. M. N. Hill) Interscience, N. Y. pp. 554–82.

Guilcher, A., 1965. Coral reefs and lagoons of Mayotte Island, Comoro Archipelago, Indian Ocean, and of New Caledonia, Pacific Ocean. In: *Submarine Geology and Geophysics*. (Eds W. F. Whittard and R. Bradshaw) Butterworths, London. pp. 21–45.

Hopley, D., 1982. *The Geomorphology of the Great Barrier Reef.* J. Wiley, Chichester. p. 453.

James, N. P. 1983. Reef. In: *Carbonate depositional Environments.* (Eds P. A. Scholle, D. G. Bebout, and C. H. Moore). Amer. Assoc. Petrol. Geol. Mem. 33. 345–462.

Jones, O. A., and Endean, R., 1973. *Biology and geology of coral reefs*, 3 Vols. Academic Press, London. 12,043 p.

Maxwell, W. G. H., 1968. *Atlas of the Great Barrier Reef.* Elsevier, Amsterdam. p. 242.

McKee, E. D., 1958. Geology of Kapingamarangi Atoll, Caroline Islands. *Bull. Geol. Soc. Amer.*, **69**, pp. 241–77.

——, Chronic, J., and Leopold, E. B., 1959. Sedimentary belts in lagoon of Kapingamarangi Atoll. *Bull. Amer. Assoc. Petrol. Geol.*, **43**, pp. 501–62.

Shepard, F. P., 1963. *Submarine Geology*, Chapter 12. Coral and other organic reefs, Harper, London. pp. 349–70.

Wiens, H. J., 1962. *Atoll environment and ecology.* Yale Univ. Press. p. 532.

Yonge, C. M., 1973. The nature of reef-building (hermatypic) corals. *Bull. Mar. Sci.*, **23**, pp. 1–15.

The account of the Permian reef complex of west Texas was based on:

Achauer, C. W., 1969. Origin of Capitan Formation, Guadalupe Mountains, New Mexico and Texas. *Bull. Amer. Assoc. Petrol. Geol.*, **53**, pp. 2314–23.

Dunham, R. J., 1969. Vadose pisolite in the Capitan Reef (Permian), New Mexico and Texas. In: Depositional environments in carbonate rocks. (Ed. G. M. Friedman) *Soc. Econ. Min. Pal.*, Sp. Pub., No. 14, pp. 182–91.

——, 1970. Stratigraphic reefs versus ecologic reefs. *Bull. Amer. Assoc. Petrol. Geol.*, **54**, pp. 1931–2.

Hills, J. M., 1968. Permian basin field area, West Texas and southeastern New Mexico. In: Saline deposits. (Ed. R. D. Mattox) *Geol. Soc. Amer.*, Sp. Pap., No. 88, pp. 17–28.

Kendall, C. G. St. C., 1969. An Environment Reinterpretation of the Permian Evaporite/Carbonate Shelf Sediments of the Guadalupe Mountains. *Bull. Geol. Soc. Amer.*, **80**, pp. 2503–26.

King, P. B., 1934. Permian stratigraphy of trans-Pecos, Texas. *Bull. Geol. Soc. Amer.* **45**, pp. 697–798.

Meissner, F. F., 1972. Cyclic sedimentation in Middle Permian strata of the Permian Basin, west Texas and New Mexico. In: *Cyclic sedimentation in the Permian Basin.* (Ed. J. C. Elam and S. Chuber) 2nd Edn. W. Texas. Geol. Soc. Midland pp. 203–32.

Newell, N. D., Rigby, J. K. Fischer. A. G., Whiteman, A. J., Hickox, J. E., and Bradbury, J. S., 1953. *The Permian Reef complex of the Guadalupe Mountains region, Texas and New Mexico.* Freeman, San Francisco. p. 236.

Rigby, J. K., 1958. Mass movements in Permian rocks of TransPecos, Texas. *J. Sediment. Petrol.*, **28**, pp. 298–315.

Silver, B. A., and Todd, R. G., 1969. Permian Cyclic Strata, Northern Midland and Delaware basins, West Texas and Southeastern New Mexico. *Bull. Amer. Assoc. Petrol. Geol.*, **53**, pp. 2223–51.

Tyrell, W. W., 1969. Criteria useful in interpreting environments of un-like but time-equivalent carbonate units (Tanshill-Capitan-Lamar), Capitan Reef Complex, West Texas and New Mexico. In: Depositional environments of carbonate rocks. (Ed. G. M. Friedman) *Soc. Econ. Min. Pal.*, Sp. Pub., No. 14, pp. 80–97.

Wilson, J. L., 1975. *Carbonate Facies in geologic history.* Springer-Verlag, New York. p. 471.

The account of the Devonian reefs of Canada was based on:

Andrichuk, J. M., 1958. Stratigraphy and facies analysis of Upper Devonian reefs in Leduc, Settler and Redwater Areas, Alberta. *Bull. Amer. Assoc. Petrol. Geol.*, **42**, pp. 1–93.

Belyea, H. C., 1960. Distribution of some reefs and banks of the Upper Devonian Woodbend and Fairholm Groups in Alberta and British Columbia. *Geol. Surv. Can., Paper, Con., Dept. Mines Tech. Survey*, 59–15. p. 7.

Cook, H. E., McDaniel, P. N., Mountjoy, E. W., and Pray, L. C., 1972. Allochthonous carbonate debris flows at Devonian bank ('Reef') margins, Alberta, Canada. *Bull. Can. Pet. Geol.*, **20**, pp. 439–97.

Evans, H., 1972. Zama—A Geophysical Case History. In: Stratigraphic oil and gas fields. (Ed. R. E. King) *Amer. Assoc. Petrol. Geol.*, Spec. Pub. No. 10, pp. 440–52.

Fischbuch, N. R., 1968. Stratigraphy, Devonian Swan Hills Reef complexes of Central Alberta. *Bull. Can. Pet. Geol.*, **16**, pp. 446–587.

Jenik, A. J., and Lerbekmo, J. F., 1968. Facies and geometry of Swan Hills Reef Member of Beaverhill Lake Formation (Upper Devonian), Goose River Field, Alberta, Canada. *Bull. Amer. Assoc. Petrol. Geol.*, **52**, pp. 21–56.

Klovan, J. E., 1964. Facies analysis of the Redwater Reef Complex, Alberta, Canada. *Bull. Can. Petrol. Geol.*, **12**, pp. 1–100.

——, 1974. Development of western Canadian Devonian reefs and comparison with holocene analogues. *Amer. Assoc. Petrol. Geol. Bull.*, **58**, pp. 787–99.

Playford, P. E., and Marathon, O. C., 1969. Devonian carbonate complexes of Alberta and Western Australia: A comparative study. *Geol. Surv. West. Aust.*, Rpt. No. 1, 43 p.

Thomas, G. E., 1962. Grouping of carbonate rocks into textural and porosity units for mapping purposes. In: *Classification of Carbonate Rocks.* (Ed. W. E. Ham) Am. Assoc. Petrol. Geol. Mem. No. 1, pp. 193–223.

The account of the Bu Hasa reef case history was based on:

Harris, T. J., Hay, J. T. C., and Twombley, B. N., 1968. Contrasting limestone reservoirs in the Murban Field, Abu Dhabi. In: *Second Regional Technical Symposium.* Soc. Pet. Eng. Dahran. 149–82.

Hassan, T. H., Mud, G. C. and B. N. Twombley. *The Stratigraphy and Sedimentation of the Thamama Group (Lower Cretaceous) of Abu Dhabi.* Ninth Arab Petroleum Congress. 107 B-3.

Twombley, B. N. and Scott, J., 1975. *Application of geological studies in the development of the Bu Hasa field, Abu Dhabi.* Ninth Arab Petroleum Congress. 133. B-1.

Other references mentioned in this chapter were:

Amstutz, G. C., and Bubinicek, L., 1967. Diagenesis in sedimentary mineral deposits. In: *Diagenesis in Sediments.* (Eds S. Larsen and G. V. Chilingar) Elsevier, Amsterdam. pp. 417–75.

——, Randohr, P., El. Baz, F., and Park, W. C., 1964. Diagenetic behaviour of sulphides. In: *Sedimentology and Ore Genesis.* (Ed. G. C. Amstutz) Elsevier, Amsterdam. p. 65.

Bathurst, R. G. C., 1975. *Carbonate sediments and their diagenesis.* Second Edition. Elsevier, Amsterdam, p. 658.

Bond, G., 1950. The Lower Carboniferous reef limestones of northern England. *J. Geol.,* **58**, 430–87.

Boulware, R. A., 1967. Seismic exploration for stratigraphic traps in Canada. *Proc. Seventh World Petrol. Cong.,* **2**, 471–80.

Braithwaite, C. J. R., 1973. Reefs: just a problem of semantics? *Bull. Amer. Assoc. Petrol. Geol.,* **57**, 1100–16.

Brown, J. S., 1968 (Ed.), *Genesis of stratiform lead-zinc-barite-fluorite deposits (Mississippi Valley type deposits).* *Economic Geology—Monograph 3.* Econ. Geol. Pub. Co. Blacksburg, Va. p. 443.

Bubb, J. N. and Hatledid, W. G., 1977. Seismic recognition of carbonate buildups. In: *Seismic Stratigraphy-applications to hydrocarbon exploration.* (C. E. Payton Ed.) Amer. Assoc. Petrol. Geol. Mem. **26**, 185–204.

Chilingar, G. V., and Terry, R. D., 1954. Relationship between porosity and chemical composition of carbonate rocks. *Petrol. Eng.,* **B-54**, pp. 341–2.

——, Bissell, H. J. and Fairbridge, R. W., (Eds). *Carbonate rocks,* Part A. Elsevier, Amsterdam. p. 471.

——, ——, and Wolf, K. H., 1967. Diagenesis of carbonate rocks. In: *Diagenesis in Sediments.* (Eds G. Larsen and G. V. Chilingar) Elsevier, Amsterdam. pp. 197–322.

——, Mannon, R. W., and Rieke, H., 1972. *Oil and gas production from carbonate rocks.* Elsevier, Amsterdam. p. 408.

Cummings, E. R., 1932. Reefs or Bioherms? *Bull. Geol. Soc. Amer.,* **43**, 331–52.

Darwin, C., 1842. *The Structure and Distribution of Coral Reefs,* London.

Davidson, C. F., 1965. A possible mode of origin of strata bound copper ores. *Econ. Geol.,* **60**, 942–54.

Derry, D. R., Clark, G. R., and Gillatt, N., 1965. The Northgate base-metal deposit at Tynagh, Co. Galway, Ireland. *Econ. Geol.,* **60**, 1218–37.

Ellison, S. P., 1955. Economic applications of Paleoecology. *Econ. Geol.* 50th Anniv., Vol. Pt. 2, pp. 867–84.

Emery, K. O., 1956. Sediments and water of Persian Gulf. *Bull. Amer. Assoc. Petrol. Geol.,* **40**, 2354–83.

——, Tracey, J. L., and Ladd, H. S., 1954. Geology of Bikini and nearby atolls. Pt. I. Geology. *U.S. Geol. Surv. Prof. Pap.* **260–A**, p. 265.

Evans, T. L., Campbell, F. A., and Krouse, H. R., 1968. A reconnaissance study of some Western Canada Pb–Zn deposits. *Econ. Geol.,* **63**, p. 349.

Ferris, C., 1972. Use of the Gravity meter in the search for Stratigraphic Traps. In: *Stratigraphic Oil and Gas Fields.* (Ed. R. E. King) Amer. Assoc. Petrol. Geol. Mem. 16. 225–243.

Fouch, T. D. and Dean, W. E., 1982. Lacustrine Environments. In: *Sandstone Depositional Environments.* (Eds P. A. Scholle and O. Spearing) Amer. Assoc. Petrol. Geol. Tulsa. 87–114.

Friedman, G. M., 1964. Early diagenesis and lithifaction in carbonate sediments. *J. Sediment. Petrol.*, **34**, 777–812.

——, 1968. Geology and geochemistry of Reefs, Carbonate Sediments and Waters, Gulf of Akaba. *J. Sediment. Petrol.*, **38**, 895–919.

Griffith, L. S., Pitcher, M. G., and Wesley-Rice, G., 1969. Quantitative environmental analysis of a Lower Cretaceous Reef Complex. In: Depositional environments in carbonate rocks. (Ed. G. M. Friedman) *Soc. Econ. Min. Pal.*, Sp. Pub., No. 14, pp. 120–38.

Guzman, E. J., 1967. Reef type stratigraphic traps in Mexico. *Proc. Seventh Petrol. Cong.*, **2**, Elsevier, Amsterdam. pp. 461–70.

Hadding, A., 1950. Silurian reefs of Gotland. *J. Geol.*, **58**, 402–9.

Harbaugh, J. W., Carbonate oil reservoir rocks. In: *Carbonate Rocks*, Part B. (Eds G. V. Chilingar, H. J. Bissell, and R. W. Fairbridge) Elsevier, Amsterdam. pp. 349–98.

Harlan Johnson, J., 1961. *Limestone-building Algae and Algal Limestones.* Colorado School of Mines. p. 297.

Heckel, P. H., 1974. Carbonate build-ups in the geologic record: a review. In: Reefs in time and space. (Ed. L. F. Laporte) *Soc. Econ. Pal. Min.*, Spec. Pub. 18, pp. 90–154.

Henson, F. R. S., 1950. Cretaceous and Tertiary reef formations and associated sediments in the Middle East. *Bull. Amer. Assoc. Petrol. Geol.*, **34**, 215–38.

Heybroek, F., 1965. The Red Sea Miocene Evaporite Basin. In: *Salt Basins Around Africa.* Inst. Pet., London. pp. 17–40.

Hills, J. M., 1968. Permian Basin field area, west Texas and Southeastern New Mexico. In: Saline Deposits. (Ed. R. M. Mattox) *Geol. Soc. Amer.*, Sp. Pap., No. 88, pp. 17–28.

Hoskin, C. M., 1966. Coral pinnacle sedimentation, Alacran Reef Lagoon, Mexico. *J. Sediment. Petrol.*, **36**, 1058–74.

Hunt, J. M., 1967. The Origin of Petroleum in Carbonate rocks. In: *Carbonate Rocks*, Part B. (Eds G. V. Chilingar, H. J. Bissell, and R. W. Fairbridge) Elsevier, Amsterdam. pp. 225–51.

Hutchison, C. S., 1983. *Economic Deposits and their Tectonic Setting.* MacMillan, London. p. 365.

Jageler, A. H., and Matuszak, D. R., 1972. Use of well logs and dipmeters in stratigraphic trap exploration. In: Stratigraphic oil and gas fields. (Ed. R. E. King) *Amer. Assoc. Petrol. Geol.*, Spec. Pub. No. 10, pp. 107–35.

Jenyon, M. K. and Taylor, J. C. M., 1983. Hydrocarbon indications associated with North Sea Zechstein shelf features. *Oil and Gas Jl.*, Dec. 5. 1983. 155–60.

Johnstone, M. H., Jones, P. J., Koop, W. J., Roberts, J., Gilbert-Tomlinson, J., Veevers, J. J., and Wells, A. T., 1967. Devonian of Western and Central Australia. In: *Intnat. Symp. Devn. System.* (Ed. D. H. Oswald) Calgary, **1**, pp. 599–612.

Kinsman, D. J., 1969. Modes of formation, sedimentary associations and diagnostic features of shallow-water and supratidal evaporites. *Bull. Amer. Assoc. Petrol.*, **53**, pp. 830–40.

Kirkland, D. W., and Evans, R., 1973. *Marine evaporites.* Benchmark papers in geology. Dowdon, Hutchinson and Ross. Stroudsburg, Penn. p. 426.

Langres, G. L. Robertson, J. O., and Chilingar, G. V., 1972. *Secondary recovery and carbonate reservoirs.* Elsevier, Amsterdam. 250 p.

Laporte, L. F. (Ed.), 1974. Reefs in time and space. *Soc. Econ. Pal. Min.*, Sp. Pub. No. 18. Tulsa. p. 255.

Leverson, A. I., 1967. *Geology of Petroleum*. Freeman, San Francisco. p. 724.

Logan, B. R., Rezak, R., and Ginsburg, R. N., 1964. Classification and environmental significance of algal stromatolites. *J. Geol.*, **72**, 68–83.

Lowenstam, H. A., 1950. Niagaran Reefs of the Great Lakes Area. *J. Geol.*, **58**, 430–87.

Manten, A. A., 1971. Silurian Reefs of Gotland. Elsevier, Amsterdam. p. 539.

Massa, D., 1965. Observations sur les series Devonniennes des confins Algero-Marocains du Sud. *C.F.P. Mem*. No. 8, Paris. p. 187.

Parkinson, D., 1944. The origin and structure of the Lower Visean reef knolls of the Clitheroe District, Lancashire. *Quart. J. Geol. Soc. London.*, **XCIX**, pp. 155–68.

Playford, P. E., and Lowry, D. C., 1966. Devonian reef complexes of the Canning Basin, Western Australia. *Geol. Surv. West Aust. Bull.*, No. 118, p. 150.

Reeckmann, A. and Friedman, G. M., 1982. *Exploration for Carbonate Petroleum Reservoirs*. J. Wiley & Sons, Chichester. p. 213.

Shearman, D. J., and Fuller, J. G. C. M., 1969. Phenomena associated with calcitization of anhydrite rocks, Winnepegosis Formation, Middle Devonian of Saskatchewan, Canada. *Proc. Geol. Soc. Lond.* No. 1969, pp. 235–9.

Sloss, L. L., 1959. Relationship of Primary Evaporites to Oil Accumulation. *Proc. Fifth Wld. Petrol. Congr*. Section 1, pp. 123–35.

Stout, J. L., 1954. Pore geometry as related to carbonate stratigraphic traps. *Bull. Amer. Assoc. Petrol. Geol.*, **48**, 329–37.

Teichert, C., 1958. Cold and deep-water coral banks. *Bull. Amer. Assoc. Petrol. Geol.*, **42**, pp. 1064–82.

Terry, C. E., and Williams, J. J., 1969. The Idris 'A' Bioherm and Oilfield, Sirte Basin, Libya—its commercial development, regional Palaeocene geological setting stratigraphy. In: *The Exploration for Petroleum in Europe and North Africa*. (Ed. P. Hepple) Inst. Petrol., London. pp. 31–48.

Toomey, D. F. (Ed.), 1981. *European Fossil Reef Models*. Soc. Econ. Pal. Min. Sp. Pub. 30. p. 546.

Weeks, L. G., 1961. Origin, migration, and occurrence of petroleum. In: *Petroleum Exploration Handbook*. (Ed. G. R. Moody) McGraw Hill, New York. p. 24.

Wilson, J. L., 1974. Characteristics of carbonate-platform margins. *Amer. Assoc. Petrol. Geol. Bull*. pp. 810–24.

# 10   Deep sea sands

## DEEP SEA SANDS DEFINED

Recent deep sea sands may be defined as those deposited below the level of the continental shelf (say about 200 m). To define what is meant by an ancient deep sea sand deposit is far more difficult. One might perhaps define an ancient deep sea sandstone as one which was laid down below effective wave base (whatever that may be, as debated on p. 185). It is extremely difficult, and generally of only academic importance, to determine palaeobathymetry accurately.

There are several criteria that may be used to conclude that a sandstone was deposited in deep water (as defined above as sub-effective wave base). Some of these criteria are negative, some positive. There should not be any features indicating subaerial exposure, or shallow water deposition. Thus a deep sea sand must obviously lack such features as desiccation cracks, rain prints, vertebrate tracks, rootlet horizons and palaeosols. Shallow water features should also be lacking, though these are more equivocal than subaerial ones. Shallow water criteria include in situ algal colonies (because algae only thrive in the photic zone), shallow water trace fossils (principally vertical burrows), and extensive cross-bedding (on the assumption that traction currents in deep seas are of insufficient velocities to generate megaripples).

The positive criteria that indicate a deep marine environment for a sand are very dubious. Examination of deep sea sediments shows that their faunal content is largely composed of transported shallow-water beasts and of pelagic free-swimming ones. Body fossils of deep marine animals are rare, though a trace fossil assemblage of invertebrate surface tracks and trails is quite common. The regional and stratigraphic position of a formation may be used to infer a deep water origin. One should suspect a deep water origin for sandstones in basin centres, especially if they are overlain by, or pass shelfward into, slope deposits (as indicated by abundant slump bedding) succeeded by shallow marine deposits. For example Walther's Law suggests a deep water origin for the Edale shales and Mam Tor sandstones of Derbyshire (refer back to Fig. 5.5). Seismic data may be useful. Sandstones encountered when drilling into the toeset of a major seismically defined prograde must have been deposited in water depths at least as great as the interval from topset to toeset (refer back to Fig. 5.16).

Despite the apparent difficulties in identifying deep sea sands most geologists accept that large volumes of ancient deep water sandstones abound in the stratigraphic record. Many of these formations were formerly referred to as 'flysch', a term which fortunately seems to be going out of fashion. Petrographically many flysch sandstones are greywackes. The sedimentary structures of flysch sandstones suggest that they are 'turbidites', having been deposited from turbidity flows, a particular type of density current. These three terms deserve definition and discussion.

*Flysch:* Thick sequences of interbedded sands and shales. The sandstones generally have erosional bases and are internally graded. The shales contain a marine fauna. First defined in the Alps this word has been applied to similar rocks in geosynclinal belts of all ages and in all parts of the world (for further data see Dzulinsky and Walton, 1965, pp. 1–12; Hsu, 1970).

*Turbidity current:* 'A current flowing on consequence of the load of sediment it is carrying and which gives it excess density' (Kuenen, 1965, p. 217).

*Turbidite:* A sediment deposited by a turbidity current.

*Greywacke:* A poorly sorted sandstone with abundant matrix, feldspar, and/or rock fragments.

For reasons shortly to be described many geologists believe that flysch sandstones are turbidites. Many flysch sandstones, but by no means all, are greywackes. Because of this the terms 'flysch', 'turbidite', and 'greywacke' have been used as synonyms. This is misleading. 'Flysch' describes a facies, 'greywacke' is a petrographic term for a particular kind of rock, and 'turbidite' is a genetic term describing the process which is *thought* to have deposited a sediment. Not only are these three words quite different but they are all extremely loosely defined in the literature.

This chapter begins with an attempt to summarize the characteristics of turbidites. This is followed by a discussion of their origin. Two case histories are then presented, one from Scotland, and one from the Alps. The chapter continues with a summary of flysch and turbidite problems and a discussion of their economic aspects. It concludes with a review of their diagnostic criteria in the sub-surface.

## DIAGNOSTIC CHARACTERISTICS OF TURBIDITES

Turbidites are identified by no single feature, but by the sum of many criteria. The selection of these is subjective and a function of the prejudices of the individual geologist. The most noticeable feature of turbidites is that they are vertically graded. Each bed generally has a sharp base and an upward gradation into shale (Fig. 10.1). Detailed examination of turbidites has shown that they contain many different types of sedimentary structure, summarized in Table 10.1. There tends to be a regular sequence of structures within individual turbidite beds (Fig. 10.2). This suite of structures has been interpreted in the following way (Walker, 1965; Harms and Fahnestock, 1965; Hubert, 1967). First a current scoured a variety of structures on a mud surface. Sedimentation of the turbidite then took place under waning current conditions. First the massive A unit was deposited, perhaps as antidunes, under an upper flow regime; shooting flow deposited the next laminated B unit and a lower flow regime deposited the micro-crosslaminated C unit. Various explanations have been offered for the upper laminated silt (Walker, 1965). Perhaps it heralds a return

*Figure 10.1* Photograph of a turbidite sandstone bed. Note the sharp base and the upper gradational contact. From the Acacus Sandstone/Tannezuft Shale transition (Silurian). Murzuk basin, Libya.

*Table 10.1. Summary of sedimentary structures associated with turbidites (see Pettijohn and Potter, 1964, for illustrations and definitions).*

| | Process (cause) | Structure (effect) |
|---|---|---|
| TIME | Erosion | Current scours, e.g. flute marks, channels, scour and fill. Obstacle scours formed down current of pebbles, shells, etc. Tool marks, i.e. groove marks due to travelling objects |
| | Deposition | Graded bedding lamination micro-crosslamination. |
| | | Convolute lamination. |
| | Deformation | Loadcasts, slumps, slides pullaparts, sand dikes, sand volcanoes. |

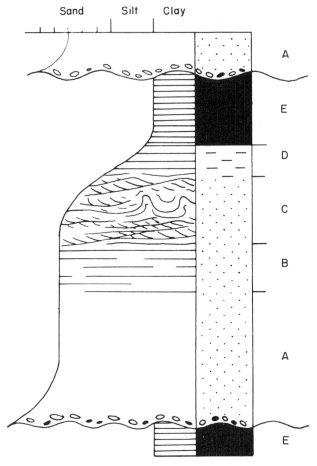

*Figure 10.2* Generalized sequence through a turbidite unit. Alphabetic letters code a vertical sequence of five different units (see text). Modified from Bouma, 1962, Fig. 8, by courtesy of Elsevier Publishing Company.

to shooting flow, but since the grain size is much finer, the actual current velocity was probably less than that which caused the lower laminated unit. The upper pelitic E unit indicates resumption of the low-energy environment which prevailed before the turbidite was laid down.

It is important to note that the sequence of structures in Table 10.1 is seldom completely developed in any one turbidite. The previous interpretation is based on the study of many turbidite sequences. Variations on this motif are discussed by Bouma (1962, pp. 49–54), Walker (1965, p. 3), and Van der Lingen (1969, p. 12). Systematic vertical variations in the frequency of bed forms within sequences of turbidites have been documented by Walker (1967, 1970). There is a tendency for only the upper units to be developed at the base of a turbidite sequence. Moving up the section, progressively lower units are present within each turbidite until complete A–E sequences appear towards the top of the series. The lower incomplete turbidites, termed 'distal', are believed to have been deposited far from the source. The upper complete units are termed 'proximal' and are believed to have been deposited near the source. A thick sequence of turbidites may thus record a gradual advance of the sediment source into the depositional area.

There has been considerable discussion about the texture of turbidites. Many ancient examples are poorly sorted with clay matrices; petrographically many are greywackes as already mentioned. However, recent deep-sea sands are widely held to be turbidites, but they are often well-sorted and clay-free (Hubert, 1964). Cummings (1962) has suggested that ancient turbidites might often have been deposited as clean, but mineralogically immature, sands in which the matrix developed from the diagenetic break-down of chemically unstable minerals. Turbidites range in grain size from silts to pebbly sandstones. The coarser varieties, often termed fluxoturbidites, are attributed to deposition by a process midway between turbidity flow and gravity slumping. Petrographically many ancient turbidites are greywackes while Recent deep-sea sands are generally protoquartzites. Protoquartzite ancient turbidites have been identified by Sturt (1961). Carbonate turbidites have been described from ancient and Recent sediments (see Table 10.2). This table also shows that turbidites have been described from a very wide range of depositional environments and are not even restricted to sedimentary rocks, but have also been described from layered gabbros (Irvine, 1965). It can be seen therefore that turbidites are characterized by the sum total of a wide range of sedimentary structures. Texturally turbidites are generally described from poorly sorted sediments, but this is not always the case. Petrographically they are most commonly recorded from greywackes, but also occur in other sandstone types and in limestones. They have been reported from a wide range of environments. With this great diversity of characteristics one might suppose that turbidites were hard to recognize. A review of the

*Table 10.2. Environments from which turbidites have been recorded.*

|  | Recent | Ancient |
|---|---|---|
| Fjords: | Holtedahl (1965) | |
| Lakes: | Grover and Howard (1938) | Kuenen (1951) |
| Deltas: | See Chapter 5 | See Chapter 5 |
| Reefs: | ? | Carozzi and Frost (1966) |
| Carbonate shelf margins: | Rusnak and Nesteroff (1964) | Thomson and Thomasson (1969); Rech-Frollo (1973) |
| Ancient flysch and Recent deep-sea sands: | See this chapter. | |
| Layered gabbros: | | Irvine (1965) |

literature suggests that, on the contrary, many geologists think that they can identify a turbidite with confidence when they see one.

From what has been said already, it is obviously important to look at modern deep sea sands when trying to understand this facies.

### Recent deep sea sands

Studies of modern continental margins show that their physiography is analogous to that of desert wadis and alluvial fans (Fig. 10.3).

The continental shelf edge is dissected by submarine channels. The source of these channels lies adjacent to either major river mouths or even coastal beaches. The channels continue down through the continental slope and debouch their sediment load onto a submarine fan, analogous to a subaerial alluvial fan. At the fan apex the submarine channel splits into a radiating complex of small channels, which are often flanked by levees. These channels become more numerous, shallower and wider down the fan. There are also changes down the fan in the sediment facies. The coarsest sand, and sometimes gravel, is deposited in the proximal (near source) part of the fan. Grain size declines distally down fan as the submarine fan sands grade out into pelagic muds of the oceanic basin floor (Shepard, 1971; Dott, 1974; Whitaker, 1976; Kelling and Stanley, 1976; Howard and Normark, 1982).

The deposits of the submarine fans show many of the characteristics of turbidites, which have been previously described and discussed.

The submarine channels are sometimes infilled by turbidites, especially at their distal ends, but they also contain the deposits of other depositional processes. At their proximal ends they may contain heterogenous slumped boulder beds, if they originate in submarine fault scarps.

If, on the other hand, the sediment in the channel is clean sand, brought in by longshore drift or from a river mouth, then a different process often prevails. This is known as grainflow and is midway between a mass flow and a turbidite (Bagnold, 1966). The sedimentary features of grain flows have been described by Stauffer (1967) and other geologists. These deposits show neither the graded bedding of turbidites nor the cross-bedding of traction deposits. They are massive

*Table 10.3. The geomorphology and deposits of deep sea sedimentary environments.*

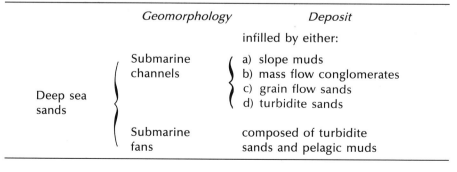

|  | *Geomorphology* | *Deposit* |
|---|---|---|
|  |  | infilled by either: |
| Deep sea sands | Submarine channels | a) slope muds<br>b) mass flow conglomerates<br>c) grain flow sands<br>d) turbidite sands |
|  | Submarine fans | composed of turbidite sands and pelagic muds |

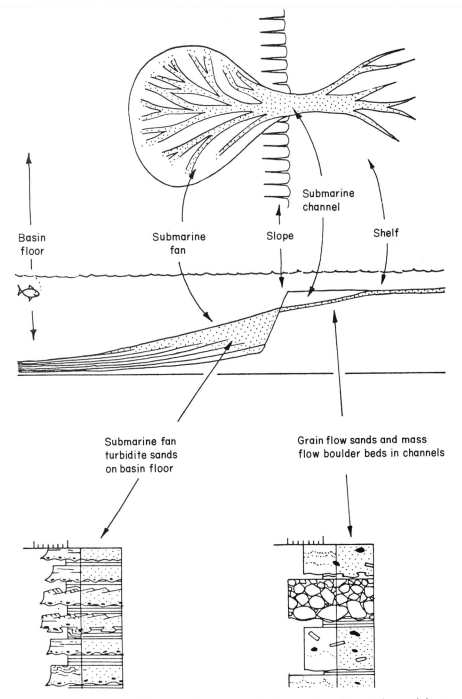

*Figure 10.3* Diagrams to illustrate the geomorphology and sedimentology of deep sea sands.

sands with abrupt upper and lower contacts. Clasts may occur scattered throughout the bed. Internally this is massive, or may show faint convoluted bedding. Sediments of this type are generally found to have been deposited on the steep slopes of sandy deltas and continental margins. They are often, but not invariably, confined to channels.

Table 10.3 shows the relationship between the geomorphology, processes and facies of submarine channels and fans.

The slopes with which these are associated may be either accretional or tectonic. Submarine channels and fans are found at the foot of modern deltas, such as the Rhône, and at the foot of such tectonic features as submarine fault scarps and continental margins. Specific accounts of such ancient analogues have been given by Mutti and Lucchi (1972) and Van de Kamp *et al.* (1974) respectively.

## DISCUSSION OF THE ORIGIN OF TURBIDITES

It has already been argued (p. 246) that the vertical fining of grain size above a scoured surface suggests that these beds were deposited from waning currents. The suggestion that this current was a turbidity flow was put forward by Kuenen and Migliorini (1950) though the germ of the idea was somewhat older (see Van de Lingen (1969) for a historical review). Subsequently the concept has been greatly elaborated and popularized. The basic idea of the turbidity flow may be summarized thus: In certain regions, such as the edges of continental shelves and deltas, rapid deposition takes place, causing the formation of a steeply sloping pile of waterlogged sediment. Every now and then the slope becomes so unstable that a triggering mechanism, such as an earthquake or storm, starts the sediment moving. Initially it slides and slumps down gullies becoming more and more liquefied until the mixture assumes the characteristics of a turbidity current. That is to say it changes into a muddy liquid whose density is greater than that of the surrounding water. It therefore moves down slope under gravity beneath the clear water. As it gains momentum it becomes erosive, cutting gullies while still on the steep slope and flute marks, etc., on reaching the open-sea floor. On this level surface the velocity starts to diminish. Deposition then takes place, first of the coarsest sediment, and, as the current wanes, of particles of finer and finer grade. When the velocity has returned to zero pelagic mud deposition continues out of suspension as before. All is quiet on the sea floor; a graded bed has been deposited on a scoured surface.

The evidence that turbidity flows deposit the sediments that geologists call turbidites is twofold. It is gathered from experiments and from observations of Recent deep-sea deposits. Graded bedding has been formed from turbidity flows generated artificially in the laboratory (Kuenen, 1948; Kuenen and Migliorini, 1950; Kuenen and Menard, 1952; and Middleton, 1966a, b, and c). Recent deep-sea sands are reported as often showing graded bedding (e.g. Shepard, 1963, p. 406; Horn *et al.*, 1971; and Horn *et al.*, 1972). The suite of internal structures found in ancient turbidites is also sometimes present (e.g. Van Straaten, 1964). The classic oft-quoted reason for believing that Recent deep-sea sands may be turbidites is the case of the 1929 Grand Banks earthquake. This had an epicentre on the edge of the continental shelf off Newfoundland. Telegraphic cables running approximately parallel to the shelf margin were broken in orderly succession downslope. Cores taken subsequently from this area showed a graded bed of silt with shallow water microfossils. This has been interpreted as the product of a turbidity flow which, generated from slumped sediment near the epicentre of the earthquake, ran down the slope ripping up cables and depositing a bed of graded sediment (Heezan and Ewing, 1952; see also Heezan, 1963, p. 743, and Shepard, 1963, p. 339). It has been widely concluded therefore that Recent deep-sea sands are the product of turbidity flows. Ancient sediments which are comparable to deep-sea sands are attributed to the same process and are categorized as turbidites.

This attractive explanation has, none the less, been criticized by a small but articulate group of geologists. Criticism is of two main types, first that deep-sea sands are not turbidites and second that ancient turbidites do not resemble Recent deep-sea sands.

The evidence against a turbidite origin for Recent deep-sea sands has been marshalled by Hubert (1964; 1967), Klein (1967), and Van de Lingen (1969). These authors point out that many deep-sea sands are clean washed, that grading is not always present, and that ripple-marking is common. Traction currents on the ocean floors are strong enough to transport sand, and ripples have been photographed at great depths (e.g. Menard, 1952). It is possible therefore that deep-sea sands are deposited not by catastrophic turbidity flows but by gentle intermittent traction currents. One of the great charms of the turbidite hypothesis was that it could explain the formation of submarine canyons on the edges of the continental shelves. These are now accessible to SCUBA divers and submersibles. Turbidity currents have not been seen in the heads of submarine canyons. Instead sand slowly creeps like a glacier, slumping and cascading over hanging valleys. Attempts to generate turbidity flows in canyon heads have been unsuccessful (Dill, 1964).

Evidence against a turbidite origin for ancient rocks which have been called turbidites hinges therefore on the two arguments that they do not resemble Recent deep-sea sands, and that these may be traction deposits anyway. Of particular interest here are the palaeocurrents of ancient turbidites. It has been noted that slump folds and turbidite bottom structures both tend to indicate movement down the palaeoslope (Walker, 1970). Sometimes however discrepancies have been noted, both between bottom structures and ripple orientations, and palaeoslope, determined from independent criteria (Kelling, 1964, and Klein, 1967). It has been suggested that the reworking of the top of a turbidite by normal oceanic currents may generate ripples migrating in directions oblique to the (?) down slope oriented bottom structures. This is an attractive compromise. However, Hubert (1967) has discovered an Alpine flysch with plant fragment orientations in the shales which coincide with the sand palaeocurrents. Since the former must be due to the normal oceanic currents perhaps the latter are too. Another curious feature is that not all flysch palaeocurrents are slope-controlled as one would expect to find in gravity-controlled turbidites. Scott (1966) found gravity-controlled slumps which had moved at right angles to associated flysch palaeocurrents. In this context it is interesting to note that Recent oceanic bottom currents are not always slope-controlled and often show a tendency to flow parallel to bathymetric contours (Klein, 1967, p. 376).

The arguments for and against a turbidite origin for ancient 'turbidites' and Recent deep-sea sands are endless. From the necessarily short preceding discussion of this problem the following points should be noted: There is a particularly widespread sedimentological phenomenon found in ancient rocks, especially flysch, which is widely interpreted as due to turbidity current deposition. This conclusion is partly based on experimental work and partly on the analogy with Recent deep-sea sands, which are believed to be turbidites. However, nobody has actually observed a turbidite in nature even in the submarine canyon heads which should be their homes. It has been argued that Recent deep-sea sands could be traction-deposited and that, anyway, they are not as similar to ancient turbidites as some authors have suggested. Ancient turbidites sometimes show features that one would not expect to find in a turbidity deposit, notably that they were laid down by currents which were not slope-controlled.

What is important is to note that Recent and ancient deep sea sands are commonly deposited where submarine channels cause sediment to be deposited on fans that prograde over basin floor muds. The submarine fan sands commonly show sedimentary features that suggest deposition from turbidity currents.

## UPPER JURASSIC DEEP SEA SANDS OF SUTHERLAND, SCOTLAND: DESCRIPTION

Mesozoic rocks of the Moray Firth basin crop out at several points along the coast of Sutherland. The most complete section is exposed in a two or three kilometres wide strip that extends for some thirty kilometres from Golspie in the south to Helmsdale in the north (Fig. 10.4). The Mesozoic strata are downfaulted against Caledonide metamorphics and intrusives together with Devonian Old Red Sandstone (Fig. 10.5). The Helmsdale fault is a branch of the Great Glen fault and marks the present boundary of the Moray Firth basin. The Helmsdale fault controlled the deposition of the Jurassic deep sea sediments to be described.

The Sutherland section begins with red fanglomerates, and alluvium of Permotriassic age. These deposits are overlain by an almost complete Jurassic section ranging from Liassic to Portlandian (Fig. 10.6). The Lower and Middle

*Figure 10.4* View looking north from Brora, Sutherland. The Helmsdale fault separates the low lying coastal strip of Jurassic sediments from the Palaeozoic basement hills to the west. The scale is provided by the Clynelish whisky distillery in the foreground.

Jurassic sediments are of a shallow water origin, including fluvio-deltaic sands, with coals, rootlet horizons, and shallow water burrows. These are interbedded with offshore shales. The first sign of deep water sedimentation appears in the early Kimmeridgian near the base of which is the Allt na Cuile Sandstone. This consists of massive multistorey channel sands with occasional scattered granules and clasts. The top of each channel is often flat-bedded (Fig. 10.7).

The Allt na Cuile Sandstone is overlain by laminated black carbonaceous shales with a sparse marine fauna and abundant plant detritus (Fig. 10.8). These beds pass gradually up into thinly interbedded shales and sandstones. The sandstones have sharp, sometimes erosional, bases and gradational tops. Internally the sandstones are massive, laminated or occasionally cross-laminated. They look like turbidites (Fig. 10.9). The laterally extensive turbidite sheet sands are locally overlain by thicker coarser massive sandstones. These are generally channelled, with intraformational shale pebble conglomerates (Fig. 10.10). The topmost 500 m or so of the exposed section consists of the Helmsdale boulder beds (Fig. 10.11). These are spectacularly large conglomerates with clasts ranging in size up to that of a highland croft. The conglomerates are interbedded with black shales and thin turbidite sands. There is an interesting vertical change in the nature of the boulder beds. Many of the clasts in the lower part are clearly reworked shallow water Jurassic sandstone. These sands are cross-bedded and occasionally contorted, suggesting that they were emplaced in a semi-consolidated state. Moving up the sequence however these clasts die out to be replaced by pebbles and boulders of Devonian Old Red Sandstone. The matrix of the boulder beds also changes upwards, becoming more calcareous, often consisting of a skeletal packstone. This is made of a diverse fauna which

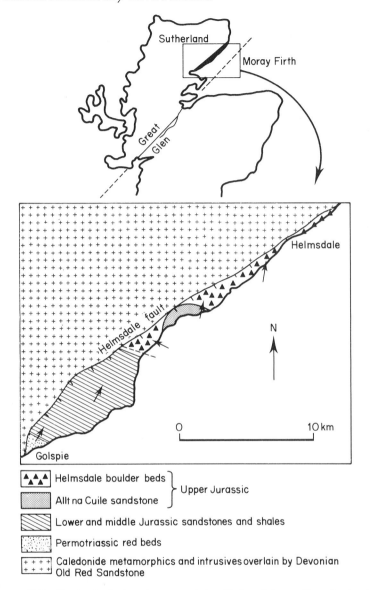

*Figure 10.5*  Geological sketch map of the Helmsdale outlier, Sutherland.

includes ammonites, brachiopods, oysters, corals, echinoids and crinoid debris. Throughout the boulder beds there is abundant slumping and occasional intrusion by sandstone dykes (Figs 10.12 and 10.13). The top of the Helmsdale boulder beds is not exposed, though shallow marine Lower Cretaceous sandstones are known to crop out on the sea floor of the Moray Firth.

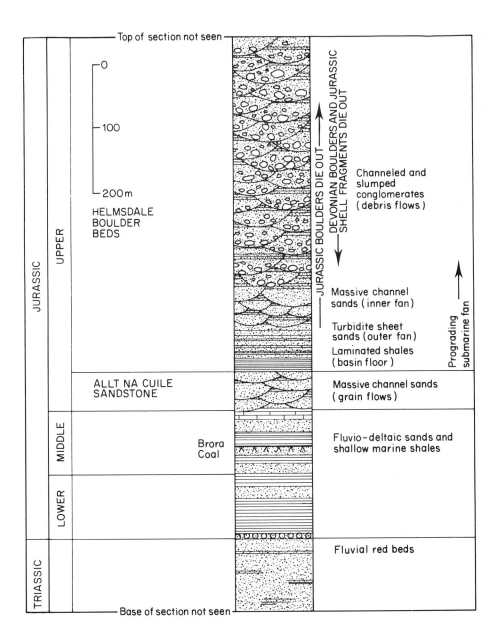

*Figure 10.6* Stratigraphic section, facies and deduced environments of the Mesozoic sediments of the Sutherland coast. Accurate formation thicknesses are only approximate due to extensive faulting. Pre-Upper Jurassic stratigraphy diagrammatic.

*Figure 10.7*  Photograph of Allt na Cuile Sandstone outcrop, showing multistorey massive channel sands, interpreted as grain flows. Lothbeg Point, Sutherland.

*Figure 10.8*  Laminated black shales of basinal marine origin. Lothbeg, Sutherland.

*Figure 10.9* Graded turbidite sands shales interpreted as outer fan deposits. Lothbeg, Sutherland.

*Figure 10.10* Massive channel sands cut into turbidite sands and shales. Interpreted as inner fan deposits. Lothbeg, Sutherland.

*Figure 10.11* Upper Jurassic Helmsdale boulder beds (fine grained variety), Lothbeg, Sutherland. Interpreted as debris flows shed from the adjacent Helmsdale fault.

## UPPER JURASSIC DEEP SEA SANDS OF SUTHERLAND, SCOTLAND: INTERPRETATION

The Permotriassic continental beds, and the paralic Lower to Middle Jurassic beds will not be considered now. It is the Upper Jurassic section that can be used to illustrate the characteristics of deep sea sands.

The Lower and Middle Jurassic sediments show the characteristics of shallow water deposition. Anomalous features are shown by the overlying massive channelled Allt na Cuile Sandstone. The absence of cross-bedding suggests that the sands may not have been deposited by traction currents. Their massive nature, together with the presence of occasional scattered clasts, suggests that they may have been deposited by channelized grainflows. This conclusion demonstrates the initiation of a slope from shallow to deeper water. There is no doubt that this slope was caused by the uplift of the Helmsdale fault.

The laminated black shales which overly the Allt na Cuile sands were clearly deposited in a sub-wave base marine environment, but the abundance of plant debris suggests that this basin lay close to the coast. The grain flow sand:basinal shale contact marks a major subsidence of the sea floor. The basinal shales are overlain by turbidite sheet sands followed by thicker pebbly channel sands. Using Walther's Law this vertical succession of facies appears to have been deposited by the progradation of a submarine fan across the basin floor. The turbidite

*Figure 10.12a and b* Slump bedding in interlaminated sands and shales within the Helmsdale boulder beds, Kintradwell, Sutherland. These features indicate deposition on a slope.

*Figure 10.13*  Sandstone dyke penetrating Upper Jurassic sediments, Kintradwell, Sutherland.

sheet sands were deposited on the outer fan and overlain by inner fan channel sands.

The overlying Helmsdale boulder beds show all the hallmarks of submarine debris flows (see Lowe, 1982). These were clearly derived from the adjacent Helmsdale fault scarp to the west. It is apparent that the gentle slope responsible

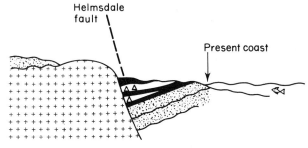

4.
? Post – Lower Cretaceous uplift. All Jurassic and much Devonian cover denuded from the upthrown block to expose the Helmsdale Granite

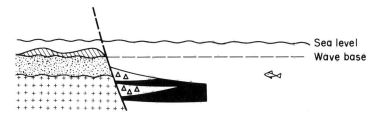

3.
Upthrown block moves above wave base. Jurassic skeletal shoal sands move across rocky Devonian substrate to form matrix of beds of Devonian boulders.

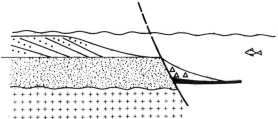

2.
Fault scarp breaks through to sea bed. Carbonaceous deltaic sand shed off as turbidites and consolidated blocks mixed with Devonian boulders

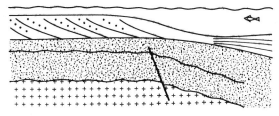

1.
Movement begins at depth on line of old Devonian Helmsdale fracture. Tilted sea bed causes grain flow slides.

*Figure 10.14* Geophantasmograms to illustrate the relationship between the Helmsdale fault and its associated Jurassic sediments (from Neves and Selley, 1975. By permission of the Norwegian Petroleum Society).

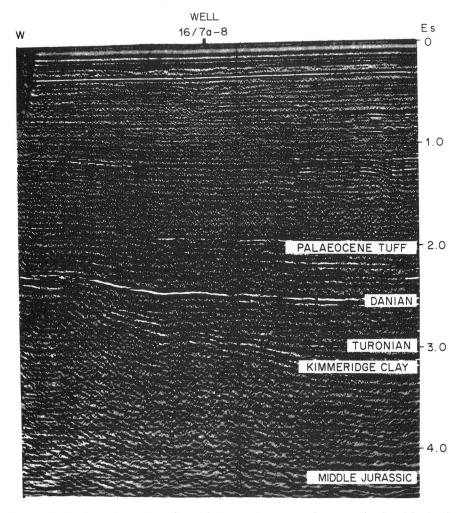

*Figure 10.15* Seismic section through Upper Jurassic submarine fan boulder beds of the Brae oil field on the margin of the Viking Graben, North Sea. This reservoir is similar to the Helmsdale boulder beds that crop out on the coast of Sutherland (from Harms *et al.* 1981, by courtesy of the Institute of Petroleum).

for the Allt na Cuile grain flows became gradually steeper, until a submarine fault scarp broke through to the surface. Initially Jurassic sediment was eroded from the upthrown block to be redistributed as turbidite sands and debris flows on the adjacent basin floor. Once this had been removed the lithified Devonian Old Red Sandstone contributed clasts to the debris flows. The hard rocky substrate of the upthrown block permitted a diverse benthonic fauna to thrive and contribute its detritus to the basinal deposits. Faulting continued after the deposition of the Upper Jurassic boulder beds. It is a curious fact that though the boulder beds are locally faulted against the Helmsdale granite they contain no clasts of it. This suggests that post-Jurassic movement may have had a transverse component. The dramatic sequence of events outlined above are illustrated in Fig. 10.14. Upper Jurassic rifting also occurred along the Viking graben of the North Sea. Many oil fields produce from Upper Jurassic deep sea sands and boulder beds that were deposited adjacent to fault scarps analogous to the one just described from Helmsdale (Magnus, Brae, Toni, Thelma, etc.). So the sedimentary model developed from the onshore outcrop is applicable to the offshore subsurface situation (Fig. 10.15).

## DESCRIPTION OF THE ANNOT TROUGH FLYSCH, ALPES MARITIMES

The second case history chosen to document this chapter is a turbidite-rich flysch series from the type area of flysch facies: the Annot trough curves around the southwestern edge of the Argentera–Mercantour massif in the Maritime Alps of southeastern France (Fig. 10.16). It contains over 650 m of flysch of Upper Eocene to Oligocene age. The flysch facies conformably overlies Upper Eocene marls in the trough axis. Laterally equivalent thick sandstone bodies in the trough margins cut into the underlying marls. Two distinct sedimentary facies can be recognized. The axial region of the trough contains the flysch. Around the edge of the trough are tongues of coarser sands and conglomerates (Fig. 10.16). The

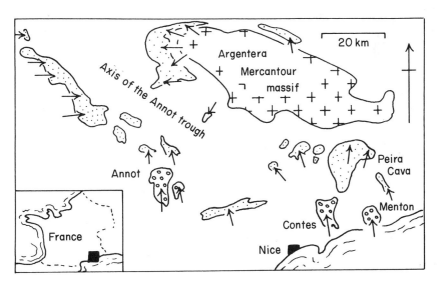

*Figure 10.16* Geological sketch map of the Annot trough, Alpes Maritimes. Submarine fan flysch facies stippled, coarse marginal submarine channel facies shown by circular ornamentation. Arrows indicate palaeocurrents. Modified from Stanley, 1967, Fig. 1.

flysch facies of the Peira Cava area have been studied in great detail. These data will now be summarized. The sequence in this area consists of the classic alternation of sandstones and shales. As the formation is traced northwards into the basin sands become less abundant and their grain size decreases. Within any vertical section the sand:shale ratio increases upward (Fig. 10.17). The sandstones are poorly sorted and range in grain size from coarse to fine. Petrographically they are hard to classify, ranging between arkose and greywacke according to the classification adopted. Sand bed thickness is variable, being generally measurable in decimetres, though beds over 1 m thick are sometimes present. Internally the sands show all the classic features of turbidites. Grading is present in nearly every bed, and the A–E sequence of intrabed units is well developed (Fig. 10.2), being first recognized and defined in this area. The base of the sandstones are erosional with a diverse suite of bottom structures including flute marks, grooves, loadcasts, and scour and fill. Palaeocurrents determined from the sandstones show a generally northerly direction of transport, diverging radially from the southern apex of the present pear-shaped outcrop of the Peira Cava formation.

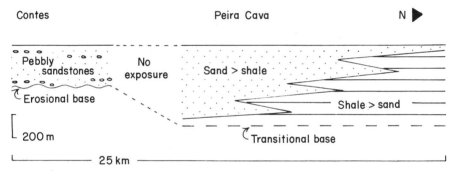

*Figure 10.17* Diagrammatic cross-section illustrating the relationship between the submarine channel of the Contes pebbly facies, and the Piera Cava Submarine fan flysch.

Trace fossils and body fossils are found in the sequence. Both the sands and the shales are sometimes burrowed, and trail covered sand:shale contacts occur. Some rare gastropods, echinoid spines, and faecal pellets have been found. There is an assemblage of pelagic foraminifera.

A second facies is developed in outliers to the south and west of the flysch in the centre of the Annot trough (Fig. 10.16). These outliers occupy north–south trending synclines. The erosional bases and restricted distribution of these patches strongly suggest that their present geometry broadly corresponds to their original distribution. They are believed to have been deposited, therefore, in huge channels up to 5 km wide and, in some cases, with a relief in excess of several hundred metres. Lithologically the sediment filling the channels contrasts markedly with the flysch to the north. Shale is rare, sandstones are coarse, poorly sorted, and pebbly. Bedding is lenticular due to channelling and there is an abundance of slump bedding. Marine shells are present with foraminifera and plant debris. Graded sand:shale units are rare. The orientations of ripples, scour marks, channel trends, and pebble imbrication indicate northerly flowing palaeocurrents.

## DISCUSSION OF THE ANNOT TROUGH SANDSTONES

The flysch sandstones of the Annot trough show all the classic features of turbidites discussed at the beginning of this chapter. The coarse marginal channel facies might, at first glance, be thought to be fluviatile. However, the presence of a marine fauna and slumps suggest that these beds were laid down in northerly sloping submarine channel complexes. The fact that these merge downcurrent with the flysch facies strongly suggests that the latter were deposited on submarine fans which built out northwards into the deeper parts of the trough. Particularly striking is the relationship between the pebbly sandstone channel complex at Contes from whose northern end the Peira Cava flysch palaeocurrents radiate (Fig. 10.16).

In conclusion this study provides a classic example of a flysch deposited to a large degree by marine slope-controlled turbidity currents. The similarity with Recent submarine canyon apron fan complexes is most striking.

## DEEP SEA SANDS: CONCLUDING DISCUSSION

This chapter has described the characteristics of a distinctive sedimentological assemblage of sedimentary structures and graded sandstones which are found in many Recent and ancient sedimentary environments. Such sequences have been widely interpreted as turbidites, due to deposition from muddy sediment suspensions of waning velocity. Evidence for ancient turbidites is based partly on experimental work and partly on observed similarities with Recent deep-sea sands interpreted as turbidites.

It is generally accepted that most deep sea sands are turbidites. Large volumes of flysch crop out today in mountain chains and are believed to have been deposited in linear troughs termed geosynclines. Many ancient flysch sandstones are poorly sorted with considerable amounts of feldspar and rock fragments. Petrographically such sandstones are greywackes.

There has been considerable discussion of the depth at which turbidites accumulate. As a general rule one might suggest that, though a turbidity flow could occur in shallow water, rapid burial or deposition below wave base would be essential to prevent the turbidite from being re-worked completely by traction currents. As we have seen, however, graded greywackes interpreted as turbidites occur in the Torridonian Series of Scotland interbedded with desiccation-cracked shales (p. 52). Stanley (1968) has reviewed a number of other shallow water turbidites.

The majority of workers, however, have favoured a deep water origin for flysch turbidites. Reasons for this conclusion are largely palaeontological. Many flysch sandstones contain fragments of shallow marine macrofossils, benthonic forams, and plant debris. The interbedded shales however contain pelagic foraminifera which may indicate considerable depths. For example, Natland and Kuenen (1951) described a 6000 m turbidite sequence from the Ventura basin, California. At the base foraminifera compare with species living today at depths of 1200–1500 m. Moving up the section foraminifera suggest shallower depths when compared with Recent assemblages. At the top of the sequence the turbidites pass into fluviatile sediments with horse bones. Such studies have been criticized on the grounds that the depth distribution of Recent foraminifera is a function not only of depth but of temperature and other ecological factors (Rech-Frollo, 1962). Two other palaeontological studies bearing on the depth of flysch should be noted. The Jaslo shales (Oligocene) of the Krosno flysch in the Polish Carpathians contain fossil fish with light-bearing organs similar to species living today at oceanic depths (Jerzmanska, 1960). Tertiary flysch from the Pyrenees contains tracks interpreted as bird footprints (Mangin, 1967). Despite this last curious case the weight of the evidence, both faunal and sedimentological, suggests that flysch sands were emplaced at depths below wave base. Deposition was either too deep or too rapid to permit the growth of an abundant benthos. The problem of the depth of deposition of flysch is discussed at greater length by Dzulinski and Walton (1965), Duff, Hallam, and Walton (1967, p. 221), and Van der Lingen (1969).

Criticisms of the turbidite hypothesis put forward to explain flysch have been based on arguments that flysch sands do not resemble deep-sea sands and that, even if they do, the latter may not be turbidites anyway, but due to waning traction currents. Support for this criticism is provided by some palaeocurrent studies which show discrepancies between the palaeoslope and the flysch palaeocurrent indicators. These should coincide if the flysch sandstones were deposited by gravity-controlled turbidity flows.

It is at present difficult to resolve the arguments for and against a turbidite origin of flysch. Despite some three decades of intensive research this fascinating facies provides as challenging a problem to the geologist as it has ever done.

## ECONOMIC ASPECTS OF FLYSCH AND TURBIDITES

Deep sea sands are obviously not potential hydrocarbon reservoirs when they have been involved in orogenesis within geosynclines. Incipient metamorphism destroys porosity, and breaks down hydrocarbons while structural deformation allows the escape of any pore fluids.

Deep sea sands in non-orogenic situations are often highly productive of oil and gas when they occur at the foot of deltas or in fault-bounded troughs with restricted marine circulation. In these situations the pelagic basin floor muds may act as source beds which generate oil and gas. The hydrocarbons can migrate updip through the turbidite fan sands with which they interfinger. Oil and gas may be trapped both in structural traps, and stratigraphically where submarine channel sands are sealed updip by impermeable slope muds. The turbidite fans generally have lower porosities and permeabilities than the grainflow channel sands, but are laterally more extensive. The Jurassic submarine fans of the North sea have already been mentioned. Palaeocene deep sea sand reservoirs produce in the Forties and Montrose fields (Parker, 1975). The Frigg field produces from a Lower Eocene submarine fan (Heritier *et al.*, 1981).

Another case of prolific oil production in turbidites occurs in western California. Here over 10 000 m of turbidites were deposited in downfaulted Tertiary basins. Subsequent tectonism has involved only gentle folding. As already mentioned (p. 268) foraminifera in the shales indicate original basin depths of some 1200 m. These shales are presumably the source rocks. Despite poor sorting and relatively low porosity and permeability the turbidite sands are good oil reservoirs. This is because individual beds are of unusually great thickness for turbidites, sometimes in excess of 3 m. Furthermore, multistorey sand sequences are common, shale between turbidites being absent either due to erosion or to a rapid frequency of turbidity flows.

Three main geometries can be recognized in Tertiary Californian turbidite facies and these can be related to deep-sea sediments now forming off the present-day Californian coast (Hand and Emery, 1964). Shoestring turbidites occupy channels sometimes located along syn-sedimentary synclines (cf. the Contes channel of the Alpes Maritimes). Sullwold (1961) cites an example 50 m thick and 600 m wide, whose associated foraminifera suggest deposition in 700 m of water. The Rosedale Channel (Miocene) of the San Joaquin basin was about one mile wide and contains some 4000 m of turbidites. It can be traced downslope for about 8 km and its fauna suggests depths greater than 400 m (Martin, 1963). Other examples are discussed by Webb (1965).

Shoestring turbidite bundles such as these probably formed in submarine channels similar to those of the present-day continental margins.

The second turbidite geometry is fan-shaped in plan and lenticular in cross-section. The Upper Miocene Tarzana fan identified by Sullwold (1961) is about 100 km wide and 1200 m thick. Associated foraminifera indicate a depositional depth of about 1000 m. Other examples are lobate and branching (Webb, 1965). Geometries such as these can be directly compared to present-day fans formed by sediment debouching from the mouths of submarine canyons on to abyssal plains.

The third facies geometry found in Californian Tertiary turbidites is a sheet. This is believed to be due to deposition on a basin floor. The Repetto Formation (Lower Pliocene) is an example of this type. It has a maximum thickness of more than 750 m and covers some 2000 km$^2$ (Shelton, 1967).

Figure 10.18 illustrates the geometry of those three types of deep sea facies. Sedimentology obviously has a role to play in the exploitation of petroleum from deep sea sands as, and when, it is found. It is important first to recognize

Turbidite infilled channels:

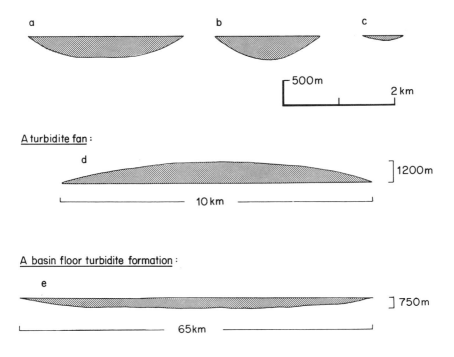

A turbidite fan:

A basin floor turbidite formation:

The examples above all drawn to the same scale with no vertical exaggeration

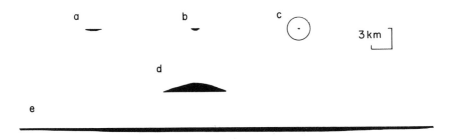

*Figure 10.18* Palaeostrike cross-sections of turbidite geometries. Sources of data cited in the text. a. Contes channel, Annot trough, Alpes Maritimes. b. Rosedale channel, San Joaquin basin, California. c. Sansina channel, Los Angeles basin, California. d. Tarzana fan, Los Angeles basin, California. e. Repetto turbidite sheet, Los Angeles basin, California.

the environment, second to identify the type of geometry, and finally to predict the trend of an individual turbidite bundle.

## SUB-SURFACE DIAGNOSIS OF DEEP SEA SANDS

Because of their economic importance great attention has been paid to the recognition of deep sea sands both at outcrop (e.g. Stanley and Unrug, 1972), and in the sub-surface (Selley, 1977; Parker, 1977; Howell and Normark, 1982; Galloway and Hobday, 1983).

The geometry of submarine channels and fans can be mapped from well logs and seismic data. Seismic lines oriented parallel to the palaeoslope may enable deep sea sands to be identified at the foot of prograding deltas or submarine fault scarps (refer back to Figs 5.5 and 10.15 respectively). Seismic sections shot parallel to the palaeostrike may show the palaeotopography of the abandoned submarine fan surface (Fig. 10.19). Sometimes it may be possible to map the complete geometry of the fan (Fig. 10.20). Note though that a fan shape is not a unique feature of submarine fans. It may also occur in alluvial fans and delta lobes.

Lithologically deep sea sediments are diverse, ranging from boulder beds to the finest silt. Glauconite shell debris, mica and carbonaceous detritus are often found mixed together in turbidite sands where the sediment is derived from both deltaic and marine shelf environments. As pointed out earlier, carbonate turbidites occur at the foot of carbonate shelves. An admixture of transported shallow water fauna in the turbidite sands with a pelagic fauna with 'Nereites' ichnofacies in the intervening shales is also typical.

When cores are available the typical Bouma A–E sequences may be found in the submarine fan turbidites, and the diagnostic features of grain flows may be found in the submarine channels.

There are particularly diagnostic gamma and S.P. log motifs but they can be confused with deltaic ones. Turbidite sequences show a very 'nervous' pattern with the curve swinging to and fro with a large amplitude. This reflects the interbedding of sands and shales, but the vertical resolution of both the gamma and S.P. logs is seldom sufficient to actually show individual graded beds. Turbidite fan sequences often show an overall upward-sanding motif which reflects the basinward progradation of the fan over basinal shales.

Submarine channels which are infilled by grainflow sands show steady featureless log profiles with abrupt lower surfaces reflecting the erosional base of the channel floor. These are commonly found above progradational fan motifs. Figure 10.21 shows these log motifs and Figure 10.22 illustrates a real example from the North Sea.

There are also characteristic dipmeter motifs in deep sea sands. Within the submarine channels the heterogenous fabric of boulder beds and grainflow sands may yield a random bag o' nails pattern. Turbidite-infilled channels may show upward-declining red patterns reflecting the gradual infilling of the curved

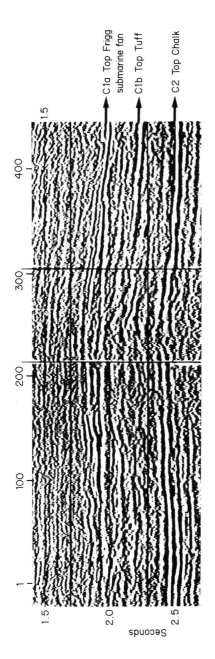

*Figure 10.19* Seismic line through the Frigg field showing the palaeotopography of the buried submarine fan reservoir. For location see Fig. 10.20. From Heritier *et al.* (1981), by courtesy of the Institute of Petroleum.

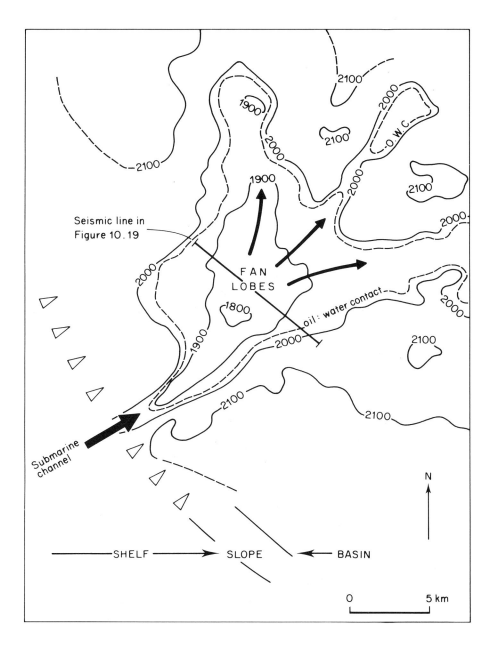

*Figure 10.20* Structure contour map (metres) on top of the Frigg oil and gas field of the North Sea. The reservoir is a submarine fan of Lower Eocene age. Based on Heritier *et al.* (1981).

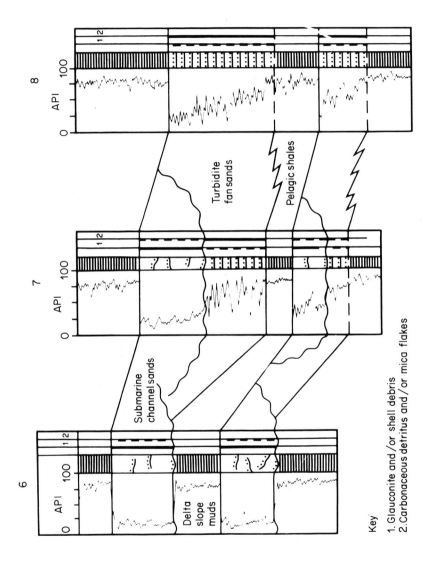

Key

1. Glauconite and / or shell debris
2. Carbonaceous detritus and / or mica flakes

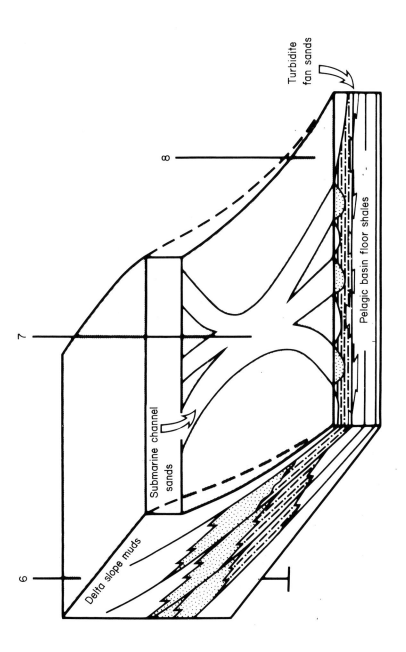

*Figure 10.21* Diagrams to illustrate the log motifs (upper) and geometry (lower) of deep sea sands. Note the similarity to the progradational and distributary channel motifs of deltas (Fig. 5.17).

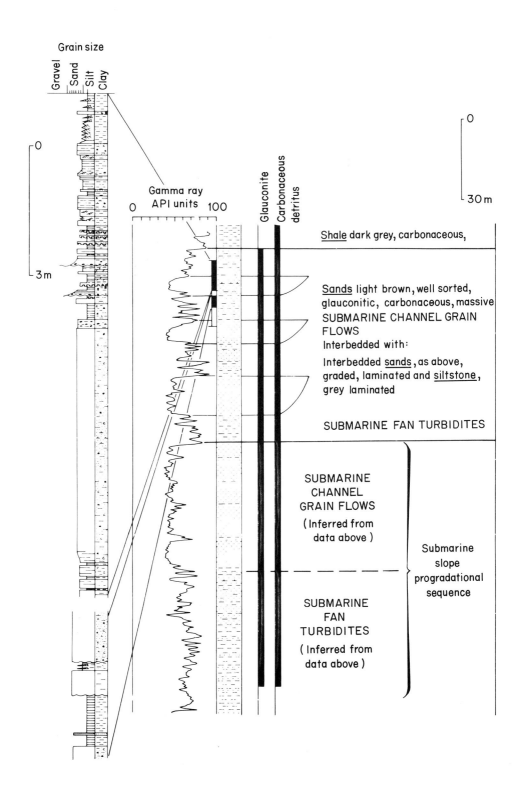

channel floor. Dips will point towards the channel axis, as in shallow water channels (see p. 137), but as cross-bedding is absent there will not be dips pointing down the channel axis. The progradational fan sequences sometimes show upward-increasing blue motifs analogous to those of delta slopes. The dips of those, as in deltas, also point in the direction of sediment progradation (Fig. 10.23).

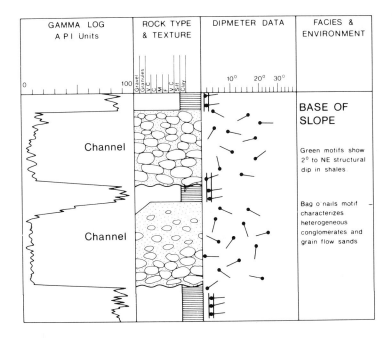

*Figure 10.23a* Deep sea sand dip motifs for slope grain flow or conglomerate channels. For explanations see text.

*Figure 10.22 (opposite page)* Sedimentary core log and gamma log of a deep sea sand from the North Sea. Note how the upper part of the core with turbidite features correlates with an erratic log curve, while the lower part of the core, interpreted as a grainflow, correlates with a steady log curve (from Selley, 1976, Fig. 6, by courtesy of the *American Association of Petroleum Geologists*).

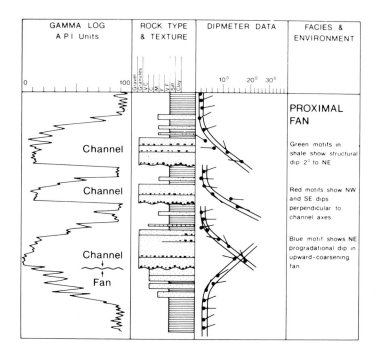

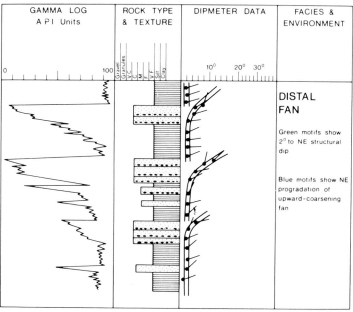

*Figure 10.23b,c* Deep sea sand dip motifs for proximal fan channel confined turbidites (b: upper) and for distal fan turbidites (c: lower). For explanations see text.

# REFERENCES

The Jurassic case history from Sutherland, Scotland, was based on:

Neves, R. and Selley, R. C., 1975. A review of the Jurassic rocks of North-east Scotland. In: *Proceedings of the Northern North Sea Symposium.* (Eds K. Finstad and R. C. Selley). Norwegian Petroleum Society, Oslo. Paper No. 5.

This paper presented the sedimentary model documented in this case history. Other relevant papers include:

Bailey, Sir E. B. and Weir, J., 1932. Submarine faulting in Kimmeridgian times: East Sutherland. *Trans. Roy. Soc. Edinb.* **47**, 431–67.

Crowell, J. C., 1961. Depositional Structures from Jurassic Boulder Beds, East Sutherland. *Trans. Edinb. Geol. Soc.* **18**, 202–19.

Lam, K. and Porter, R., 1977. The distribution of palynomorphs in the Jurassic rocks of the Brora Outlier. NE Scotland. *Jl. Geol. Soc. Lond.* **134**, 45–55.

Pickering, K. T., 1984. The Upper Jurassic 'Boulder Beds' and related deposits: a fault-controlled submarine slope, NE Scotland. *Jl. Geol. Soc. Lond.* **141**, 357–74.

The description of the Annot Trough Flysch was based on:

Bouma, A. H., 1959a. Flysch Oligocéne de Peira-Cava (Alpes Maritimes, France). *Eclogae Geol. Helv.*, **51**, 893–900.

——, 1959b. Some data on turbidites from the Alpes Maritimes (France). *Geol. Mijnb.*, **21**, 223–7.

——, 1962. *Sedimentology of Some Flysch Deposits.* Elsevier, Amsterdam. p. 168.

Stanley, D. J., 1962. Etudes sédimentologiques des grès d'Annot et de leurs équivalent latéraux. *Soc. Eds. Technip.* Paris. Ref. 6821, p. 158.

——, 1963. Vertical petrographic variability in Annot sandstones turbidites: some preliminary observations and generalizations. *J. Sediment. Petrol.*, **33**, 783–8.

——, 1965. Heavy minerals and provenance of sands in flysch of central and southern French Alps. *Bull. Amer. Assoc. Petrol. Geol.*, **49**, 22–40.

——, 1967. Comparing patterns of sedimentation in some ancient and modern submarine canyons. *Preprint Seventh Internat. Sedol. Congress*, 4 p.

——, 1967. Comparing patterns of sedimentation in some modern and ancient submarine canyons. *Earth and Planetary Science Letters*, **3**, pp. 371–80.

——, and Bouma, A. H., 1964. Methodology and palaeogeographic interpretation of flysch formations: a summary of studies in the Maritime Alps. In: *Turbidites.* (Eds A. H. Bouma and A. Brouwer) Elsevier, Amsterdam, pp. 34–64.

Other references cited in this chapter:

Bagnold, R. A., 1966. An approach to the sediment transport problem from general physics. *U.S. Geol. Surv. Prof. Paps.*, 422–I. p. 37.

Bouma, A. H., and Brouwer, A., 1964. *Turbidites.* Elsevier, Amsterdam. p. 264.

Cummings, W. A., 1962. The greywacke problem. *Lpool. Manchr. Geol. J.*, **3**, 51–72.

Carozzi, A. V., and Frost, S. H., 1966. Turbidites in dolomitized flank beds of Niagaran (Silurian) Reefs, Lapel, Indiana. *J. Sediment. Petrol.*, **36**, 563–75.

Dill, R. F., 1964. Sedimentation and erosion in Scripps submarine canyon head. In: *Marine Geology.* (Ed. R. L. Miller) Macmillan, London. pp. 23–41.

Dott, R. H. (Ed.), 1974. Modern and Ancient geosynclinal sedimentation. *Soc. Econ. Pal. Min.*, Spec. Pub. No. 9.

Duff, P. McL. D., Hallam, A., and Walton, E. K., 1967. *Cyclic Sedimentation.* Elsevier, Amsterdam. p. 280.

Dzulinski, S., and Walton, E. K., 1965. *Sedimentary Features of Flysch and Greywackes.* Elsevier, Amsterdam. p. 274.

Galloway, W. E. and Hobday, D. K., 1984. *Terrigenous Clastic Depositional Systems.* Springer-Verlag, Berlin. p. 423.

Grover, N. C., and Howard, C. S., 1938. The passage of turbid water through Lake Mead. *Trans. Amer. Soc. Civ. Eng.*, **103**, 720–90.

Hand, B. M., and Emery, J. O., 1964. Turbidites and topography of North end of San Diego trough, California. *J. Geol.*, **72**, pp. 526–42.

Harms, J. C., and Fahnestock, R. K., 1965. Stratification, bed forms, and flow phenomena (with an example from the Rio Grande). In: Primary sedimentary structures and their hydrodynamic interpretation. (Ed. G. V. Middleton) *Soc. Econ. Min. Pal.*, Sp. Pub., No. 13, pp. 84–115.

——, Tackenberg, P., Pickles, E. and Pollock, R. E., 1981. The Brae oilfield area. In: *Petroleum Geology of the Continental Shelf of North-west Europe.* (Eds L.V. Illing and G. D. Hobson). Heyden Press, London. 352–57.

Heritier, F. E., Lossel, P., and Wathne, E., 1981. The Frigg gas field. In: *Petroleum Geology of the Continental Shelf of North-West Europe.* (Eds L. V. Illing and G. D. Hobson). Heyden Press, London. 380–94.

Heezen, B. C., 1963. Turbidity Currents. In: *The Sea*, vol. III. (Ed. M. N. Hill) Interscience, N.Y. pp. 742–75.

——, and Ewing, M., 1952. Turbidity currents and submarine slumps and the Grand Banks earthquake. *Am. J. Sc.*, **250**, 849–73.

Holtedahl, H., 1965. Recent turbidites in the Hardangerfjord, Norway. In: *Submarine Geology and Geophysics.* (Eds W. F. Whittard and R. Bradshaw) Butterworths, London. pp. 107–42.

Horn, D. R., Ewing, M., Delach, M. N., and Horn, B. M., 1971. Turbidites of the northeast Pacific. *Sedimentology*, **16**, 55–69.

——, Ewing, J. I., and Ewing, M., 1972. Graded-bed sequences emplaced by turbidity currents north of 20° in the Pacific, Atlantic and Mediterranean. *Sedimentology*, **18**, 247–75.

Howell, D. G. and Normark, W. R., 1982. Submarine Fans. In: *Sandstone Depositional Environments.* (Eds P. A. Scholle and D. Spearing). Amer. Assoc. petrol. Geol. Tulsa. 365–404.

Hsu, K. J., 1970. The meaning of the word Flysch, a short historical search. In: Flysch sedimentology in North America. (Ed. J. Lejoie) *Geol. Soc. Can.*, Sp. Paper 7. pp. 1–11.

Hubert, J. F., 1964. Textural evidence for deposition for many western North Atlantic deep-sea sands by ocean bottom currents rather than turbidity currents. *J. Geol.*, **72**, 757–85.

——, 1967. Sedimentology of pre-alpine flysch sequences, Switzerland. *J. Sediment. Petrol.*, **37**, 885–907.

Irvine, T. N., 1965. Sedimentary structures in Igneous intrusions with particular reference to Duke Island Ultramafic Complex. In: Primary sedimentary structures and their hydrodynamic interpretation. (Ed. G. V. Middleton) *Soc. Econ. Min. Pal.*, Sp. Pub., No. 13, pp. 220–32.

Jerzmanska, A., 1960. Ichthyo-fauna from the Jaslo shales at Sobniow (Poland). *Acta. Palaeontol. Polon.*, **5**, 367–419.

Kelling, G., 1964. The turbidite concept in Britain. In: *Turbidites.* (Eds A. H. Bouma and A. Brouwer) Elsevier, Amsterdam. pp. 75–92.

——, and Stanley, D. J., 1976. Sedimentation in canyon, slope, and base-of-slope environments. In: *Marine sediment transport and environment management.* (Eds. D. J. Stanley and D. J. P. Swift) J. Wiley, New York. pp. 379–435.

Klein, G. de V., 1967. Paleocurrent analysis in relation to modern sediment dispersal patterns. *Bull. Amer. Assoc. Petrol. Geol.*, **51**, pp. 366–82.

Kuenen, P. H., 1948. Turbidity currents of high density. *Int. Geol. Congr.* 18th Session, pp. 44–52.

——, 1951. Mechanics of varve formation and the action of turbidity currents. *Geol. Fören, Stockholm Förh.*, **73**, 69–84.

——, 1965. Comment. In: Primary sedimentary structures and their hydrodynamic interpretation. (Ed. G. V. Middleton) *Soc. Econ. Min. Pal.*, Sp. Pub., No. 13, pp. 217–18.

——, 1967. Emplacement of flysch-type sand beds. *Sedimentology*, 9, 203–43.

——, and Migliorini, C. I., 1950. Turbidity currents as a cause of graded bedding. *J. Geol.*, **58**, 91–127.

Lajoie, J., 1970. Flysch sedimentology in North America. *Geol. Soc. Can.*, Sp. Pap. 7. p. 272.

Lowe, D. R., 1982. Sediment gravity flows: II. Depositional models with special reference to the deposits of high-density turbidity currents. *J. Sediment. Petrol.* **52**, 279–97.

Mangin, J. P., 1962. Traces de pattes d'oiseaux et flute-casts associés dans un 'facies flysch' du Tertiare pyrénéen. *Sedimentology*, **1**, 163–66.

Martin, B. D., 1963. Rosedale Channel: Evidence for Late Miocene submarine erosion in Great Valley of California. *Bull. Amer. Assoc. Petrol. Geol.*, **47**, pp. 441–56.

Menard, H. W., 1952. Deep ripple marks in the sea. *J. Sediment. Petrol.*, **33**, 3–9.

Middleton, G. V., 1966a. Small-scale models of turbidity currents and the criterion for auto-suspension. *J. Sediment. Petrol.*, **36**, 202–8.

——, 1966b. Experiments on density and turbidity currents. I. Motion of the head. *Can. Jnl. Earth Sc.*, **3**, 523–46.

——, 1966c. Experiments on density and turbidity currents. II. Uniform flow of density currents. *Can. Jnl. Earth Sc.*, **3**, 627–37.

——, 1967. Experiments on density and turbidity currents. III. *Can. Jnl. Earth Sc.*, **4**, 475–505.

Mutti, E. and Lucchi, F. R., 1972. Le turbiditii dell'Appennino settentrionale: introduzione all'analisi di facies. *Mem. Soc. Geol. Ital.*, **XI**, pp. 161–99.

Natland, M. L., and Kuenen, P. H., 1951. Sedimentary history of the Ventura Basin, California, and the action of turbidity currents. *Soc. Econ. Min. Pal.*, Sp. Pub., No. 2, pp. 76–107.

Parker, J. R., 1975. Lower Tertiary sand development in the central North Sea. In: *Petroleum and the continental shelf of northwest Europe.* (Ed. A. W. Woodland) Applied Science Publishers, London. pp. 447–56.

——, 1977. Deep-Sea Sands. In: *Developments in Petroleum Geology*—1. G. D. Hobson. (Ed.). Applied Science Pubs. pp. 225–242.

Pettijohn, F. J., and Potter, P. E., 1964. *Atlas and glossary of primary sedimentary structures.* Springer-Verlag, Berlin. p. 370.

Rech-Frollo, M., 1962. Quelques aspects des conditions de dépôt du flysch. *Bull. Geol. Soc. France*, **7**, pp. 41–8.

——, 1973. Flysch carbonates. *Bull. Centre Rech. Pan.*, **7**, pp. 245–63.

Rusnak, G. A., and Nesteroff, W. D., 1964. Modern turbidites: terrigenous abyssal plain versus bioclastic basin. In: *Marine Geology.* (Ed. R. L. Miller) Macmillan, New York. pp. 488–503.

Scott, K. M., 1966. Sedimentology and dispersal pattern of a Cretaceous flysch sequence, Patagonian Andes, southern Chile. *Bull. Amer. Assoc. Petrol. Geol.*, **50**, pp. 72–107.

Selley, R. C., 1976. Sub-surface environmental analysis of North Sea sediments. *Bull. Amer. Assoc. Petrol. Geol.*, **60**, 184–95.

——, 1979. Dipmeter and log motifs in North Sea submarine-fans. *Bull. Amer. Assoc. Petrol. Geol.* **63**, 905–917.

Shelton, J. W., 1967. Stratigraphic models and general criteria for recognition of alluvial, barrier bar and turbidity current sand deposits. *Bull. Amer. Assoc. Petrol. Geol.* **51**, pp. 2441–60.

Shepard, F., 1971. *Submarine canyons and other sea valleys*. Wiley, New York. p. 381.

Stanley, D. J., 1968. Graded bedding—sole markings—graywacke assemblage and related sedimentary structures in some Carboniferous flood deposits, eastern Massachusetts. *Geol. Soc. Amer. Spec. Pap.* No. 106, pp. 211–39.

——, and Unrug, R., 1972. Submarine channel deposits, fluxoturbidites and other indicators of slope and base of slope environments in modern and ancient marine basins. In: Recognition of Ancient sedimentary environments. (Eds. J. K. Rigby and W. K. Hamblin) *Soc. Econ. Pal. Min.*, Sp. Pub. No. 16. pp. 287–340.

Stauffer, P. H., 1967. Grainflow deposits and their implications, Santa Ynez Mountains, California. *J. Sediment. Petrol.*, **37**, 487–508.

Sturt, B. A., 1961. Discussion in: Some aspects of sedimentation in orogenic belts. *Proc. Geol. Soc. Land.*, No. 1587, p. 78.

Sullwold, H. H., 1961. Turbidites in oil exploration. In: *Geometry of Sandstone Bodies*. (Eds. J. A. Peterson and J. C. Osmond) Amer. Assoc. Petrol. Geol. pp. 63–81.

Thomson, A. F., and Thomasson, M. R., 1969. Shallow to deep water facies development in the Dimple limestone (Lower Pennsylvanian), Marathon Region, Texas. In: Depositional environments in carbonate rocks. (Ed. G. M. Friedman) *Soc. Econ. Min. Pal.*, Sp. Pub., No. 14, pp. 57–78.

Van de Kamp, P. C., Conniff, J. J., and Morris, D. A., 1974. Facies relations in the Eocene-Oligocene in the Santa Ynez Mountains, California. *J. Geol. Soc. Lond.*, **130**, 554–566.

Van der Lingen, G. J., 1969. The turbidite problem. *N. Z. Jl. Geol. Geophys.*, **12**, 7–50.

Van Straaten, L. M. J. U., 1964. Turbidite sediments in the southeastern Adriatic sea. In: *Turbidites*. (Eds A. H. Bouma and A. Brouwer) Elsevier, Amsterdam. pp. 142–7.

Walker, R. G., 1965. The origin and significance of the internal sedimentary structures of turbidites. *Proc. Yorks. Geol. Soc.*, **35**, pp. 1–32.

——, 1967. Turbidite sedimentary structures and their relationship to proximal and distal depositional environments. *J. Sediment. Petrol.*, **37**, 25–43.

——, 1970. Review of the geometry and facies organisation of turbidites and turbidite-bearing basins. In: Flysch sedimentology of North America (Ed. J. Lajoie) *Geol. Soc. Can.*, Sp. Pub. pp. 219–51.

Webb, F. W., 1965. The stratigraphy and sedimentary petrology of Miocene turbidites in San Joaquin Valley (Abs.), *Bull. Amer. Assoc. Petrol. Geol.*, **49**, p. 362.

Whitaker, J. H. McD. (Ed.), 1976. *Submarine canyons and deep-sea fans. Modern and Ancient*. Dowden, Hutchinson and Ross, Stroudsburg, Pa. p. 426.

# 11 Pelagic deposits

## INTRODUCTION:
## RECENT PELAGIC DEPOSITS

This chapter describes pelagic deposits. The term pelagic, 'is generally applied to marine sediments in which the fraction derived from the continents indicates deposition from a dilute mineral suspension distributed throughout deep-ocean water' (Arrhenius, 1963, p. 655). To prove that an ancient sediment was deposited in a pelagic environment, as defined above, it is necessary to demonstrate both great depth and an absence of terrestrial influence. As this chapter shows it is not easy to prove both these points in ancient sediments. It may be safer therefore to define pelagic deposits as 'of the open sea' (Riedel, 1963, p. 866).

The need to make this distinction between depth and 'oceanicity' in analysing the environment of supposed ancient pelagic deposits has been stressed by Hallam (1967, p. 330). The Recent seas can be divided into those of the continental shelves which are separated by the shelf break, at about 200 m from the oceans.

Deposits of the Recent oceans have been studied intensively. Particularly important accounts of the modern ocean floor sediments are given in the Deep Sea Drilling Project (D.S.D.P.) reports. More concise reviews occur in volumes edited by Lisitzin (1972), Inderbitzen (1974), Hsu and Jenkyns (1974), Cook and Enos (1977), Ross (1982), Scrutton and Talwani (1982), and Stow and Piper (1984).

Recent oceanic sediments can be broadly classified into these types:

(1) Terrigenous sediments,
(2) calcareous oozes,
(3) siliceous oozes,
(4) red clays,
(5) manganiferous deposits.

These will now be briefly described and their distribution discussed.

Terrigenous sediments are present adjacent to the continents. They consist of argillaceous deposits and the deep-sea sands, of possible turbidite origin, discussed in the previous chapter. Calcareous oozes, or muds, are composed largely of the shells of microfossils (Lucas, 1973). Two main types may be distinguished: Pteropod ooze made up largely of the aragonitic shells of this mollusc, and foraminiferal oozes, composed largely of calcitic foraminiferal tests, often of *Globigerina*.

Siliceous oozes are made of the skeletons of Diatoms and Radiolaria.

Red clays are red and dark brown muds which are believed to form from the finest of wind-blown dust from continental deserts, together with extremely fine particles of volcanic ash and cosmic detritus.

The last type of pelagic deposit is diagenetic rather than depositional in origin; these are scoured sea floor surfaces termed 'hard grounds' impregnated and overlain by nodules rich in manganese. It has been estimated that this is the most extensive type of hard rock surface on the lithosphere (Mero, in Shepard, 1963, p. 404).

The distribution of the various Recent pelagic deposits is roughly correlated with depth. Thus over much of the Atlantic and Pacific oceans Red clays occur in the deepest parts, Radiolarian oozes form in shallower waters deeper than about 4500 m, calcitic foraminiferal oozes lie on the ocean bed between 4500 m and 3500 m, above which point aragonitic pteropod and foraminiferal oozes are found (Fig. 11.1). Though the distribution of pelagic deposits is loosely correlated with depth it is actually controlled by a variety of factors such as rate of sedimentation and rate of solution. Thus the sequence, with increasing depth, of aragonitic, calcitic, siliceous, and argillaceous deposits largely reflects their increasing chemical stability. The rate of solution of these minerals is a function of their rate of burial, the water temperature and its state of saturation by the various chemicals, as well as the hydrostatic pressure. Only the last of these is actually depth-dependent. This point is illustrated by the fact that, regardless of depth, calcareous oozes are poorly developed in polar regions. This is due to the low temperature of the bottom water which causes the solution of aragonite and calcite at faster rates than in the more equable equatorial oceans.

This fact, that the distribution of Recent pelagic deposits is not a direct reflection of depth, is critical to attempts to determine absolute depths of ancient pelagic deposits.

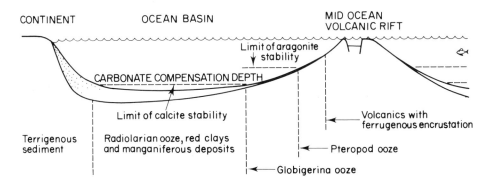

*Figure 11.1* Diagrammatic cross-section through a modern ocean basin, complete with mid-ocean rift, to show the relationship between deposits and water depth. Note that the carbonate compensation depth varies with latitude.

At the present time the oceans cover much more of the earth's surface than the continental regions. There does not appear to be a corresponding dominance of deep over shallow water sediments in the geological column; on the contrary, ancient deep-sea deposits appear to be very rare. This supports the concept of isostasy since it suggests that the continents have never been submerged to great depths below the oceans. Not only are ancient deep-sea deposits very rare but their depth is often debatable. This is because it is very hard to separate criteria of depth of deposition from those which indicate distance from the land. The two are not synonymous as the wide continental shelves of the Atlantic indicate today. This dilemma will become apparent in the discussion of the case history which follows.

## TRIASSIC-JURASSIC DEPOSITS OF THE MEDITERRANEAN: DESCRIPTION

Lengthy geological studies have revealed that during Mesozoic and early Tertiary time marine sediments were laid down in a trough stretching from southern France, northern Italy, and beyond to the east. In mid-Tertiary time these deposits were folded and uplifted to form the Alps and associated mountain chains. This trough is called the Tethys geosyncline. In general the following sequence of events seems to have taken place in this region:

(1) Deposition of marine carbonates and fine-grained clastics (Triassic-Jurassic); thought to be shelf sediments;

(2) Deposition of thin sequences of clays, cherts, and limestones, sometimes associated with vulcanicity (Triassic-Lower Cretaceous); thought to be pelagic in origin;

(3) Deposition of thick flysch sand: shale sequences Cretaceous-Oligocene); thought to be turbidites;

(4) Phase of folding, thrusting and uplift;

(5) Deposition of clastic sediments in troughs marginal to the mountains, e.g. the Swiss plain and the north Italian plain. This, the molasse facies, is generally continental, at its base is locally marine and sometimes calcareous (Oligocene-Recent).

The timing of these events varies considerably from place to place; two phases often occurring simultaneously in adjacent areas. The clays, cherts and limestones of the second phase are often attributed to deposition in pelagic and deep water environments. They will now be described and the reasons for this interpretation will be discussed. These deposits may be divided into three sub-facies defined as follows:

(1) Clay shales, marls, and micritic limestones,

(2) radiolarian cherts,

(3) nodular red limestones.

## Clay shales, marls, and micritic limestones

This sub-facies consists of clay shales, sometimes dark and pyritic, sometimes calcareous, marls, and fine-grained micro-crystalline limestones. These beds are generally massive or laminated. The various lithologies sometimes occur rhythmically interbedded with one another. Fossils are generally rare but often well-preserved and sometimes crowded on bedding surfaces. They include thin-shelled lamellibranchs termed *Posidonia* or *Bositra*, belemnites, ammonites, and brachiopods. Often ammonite shells are absent but the lids of their shells (aptychi) occur. A particularly characteristic brachiopod of this facies is *Pygope*, an unusual form with a hole in the middle. Beds of this sub-facies are named according to their fauna, e.g. Aptychus marls, Pygope limestone, and so on. It is widely distributed in Upper Jurassic and Lower Cretaceous rocks of the Alps, the Apennines, and Greece.

## Radiolarian cherts

Radiolarian cherts are often associated with the above sub-facies and share a similar geographic and stratigraphic distribution. They are generally thin-bedded and dark grey, black, or red in colour. Bedding is often rhythmic with thin shale partings. Siliceous limestones are sometimes present transitional between the cherts and the micrites of the previous sub-facies. The cherts are often largely composed of the microscopic siliceous tests of Radiolaria. In addition they contain minor quantities of silicified sponge spicules and foraminifera. The macrofauna is similar to that of the previous sub-facies but is generally much rarer. This chert facies has been considered as an insoluble residue from which all calcium carbonate has been dissolved.

## Nodular red limestones

Red nodular limestones, termed 'ammonitico rosso', occur associated with the two previous sub-facies and are widespread over the Alps, Apennines, and parts of Greece. They range in age from Triassic to Jurassic. Lithologically this sub-facies consists of layers of pink micrite nodules tightly packed in a dark red marl matrix. The nodules are often coated with iron and manganese (Fig. 11.2). Thin beds of manganese nodules are sometimes present. The characteristic fossils of this facies are the ammonites from which it gets its name. These are locally abundant. Also present are the aptychi of ammonites, belemnites, crinoids, echinoids, lamellibranchs, gastropods, foraminifera, and sponge spicules. These fossils often show signs of corrosion and mineral impregnation. Scoured bedding surfaces, termed 'hard grounds', are sometimes present. Hard ground substrates are generally very rich in iron and manganese and are occasionally phosphatized. They are sometimes pierced by *Chondrites* burrows and overlain by a thin layer of manganese or phosphate nodules.

*Figure 11.2* Polished slab of Ammonitico Rosso, Adnet Limestone, Lower Lias, Adnet, Austria. Pale pink micritic limestone fragments set in a matrix of red clay rich in iron and manganese. From Hallam (1967 Plate 2. Fig 1.) by courtesy of the Editors of the Scottish Journal of Geology.

## TRIASSIC-JURASSIC DEPOSITS OF THE MEDITERRANEAN: DISCUSSION

The three sub-facies just described have a number of points in common. They all show an absence of land-derived quartzose sediment and have obviously marine faunas. Due to complex tectonics, their stratigraphic relations are often uncertain, but palaeontology indicates that they were deposited at the same time as adjacent thicker shelf sediments whose depth must be relatively shallow since they are sometimes reefal. These rocks are often overlain by flysch turbidites. All the sub-facies are fine-grained and devoid of sedimentary structures such as cross-bedding and channelling which would suggest strong current action. The 'hard grounds' may be due to gentle scouring of early cemented sea beds. The fauna is characterized by an abundance of pelagic fossils and a scarcity of benthos. Though the shells are generally preserved as calcite, as are most fossils, there is a curious absence of shells of original aragonitic composition. Thus the calcitic aptychi of ammonites often occur while their aragonitic shells do not. Chemically the rocks are unusual due to the abundance of organic silica and calcite, together with the presence of unusual amounts of iron, phosphate, and manganese.

The fine grain size and absence of cross-bedding and channelling indicate that these are low-energy deposits which must have formed below wave base, though as the 'hard grounds' show, subjected to intermittent scouring by currents.

According to the deductions just made these deposits are of pelagic origin, in the sense that they originated in the open sea far from continental influence. There is, though, some controversy as to the precise depth of their deposition. The evidence cited in favour of great depth hinges on the analogy of these sediments with Recent deep-sea deposits. Similarities are to be found in the absence of land-derived clastics, the scarcity of current structures, and scarcity of benthos, despite the frequent oxygenation of the sea bed, indicated by the presence of red ferric iron. The limestones invite comparison with Recent Globigerina oozes, the cherts with Recent radiolarian oozes, the manganiferous beds with the red clays and manganiferous nodules. The lack of fossils of aragonitic shells can be attributed to the greater solubility of aragonite over calcite. At the present day aragonite goes into solution at depths of about 3500 m (Friedman, 1965) and calcite at about 4500 m (Berner, 1965). Hard ground horizons occur between 200 and 3500 m (Fischer and Garrison, 1967).

All these analogies suggest a deep water origin for the Alpine beds described above. On the other hand it can be argued that many of these criteria are either dubious on chemical grounds, or that they indicate distance from the land, not great depth. This is particularly valid in the case of absence of land-derived quartz and the abundance of fine-grained clay, and organic calcareous and siliceous sediment. Geochemical studies of Recent deep-sea nodules have laid stress on their oxygen, iron, and manganese ratios and on their trace element content. Recent manganese nodules occur today in widely varying depths and even in some lakes. Their chemical variation may be not just depth-controlled but also due to proximity of land-derived chemicals (Price, 1967). Similarly the solution rates of aragonite and calcite are controlled not just by depth, but also by temperature, as pointed out in the introduction to this chapter. Sedimentation rate will also control solution, slow burial increasing the time available for a shell to corrode on the sea bed (Hudson, 1967).

The facies relationships of the supposed deep-water deposits are sometimes curious too. For example in the Unken syncline of Austria there is a 300-m sequence of Jurassic pelagic limestones, radiolarian cherts, and ammonitico rosso for which Garrison and Fischer (1969) have discussed depths in the order of up to 4000–5000 m. This sequence overlies a karstic topography of Upper Triassic limestones and at the top is unconformably overlain by Lower Cretaceous (Berriasian) limestones. There must have been quite a tectonic flutter between the end of the Triassic and the Lower Cretaceous.

In conclusion the fine-grained limestones, ammonitico rosso, and radiolarian cherts of the Mediterranean are clearly marine deposits. They were laid down in a low-energy environment below the level of strong current activity and, as

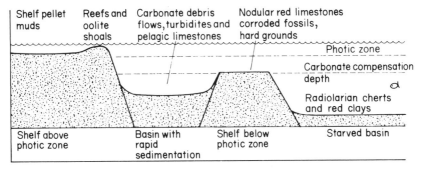

*Figure 11.3* Geophantasmogram to show the formation of Ammonitico Rosso and related facies of the Tethys during the Triassic-Jurassic periods. Note that fault blocks may yoyo from the photic zone to below the carbonate compensation depth in response to tectonic and eustatic events. (Simplified from Bernoulli and Jenkyns, 1974).

the absence of algae show, below the photic zone. The condensed stratigraphy shows that sedimentation was slow, with little influx of land-derived material (Fig. 11.3). They may be attributed to a pelagic environment with confidence. Comparison with Recent deep sea deposits suggests a deep-water origin. However the distribution of these is not directly depth-dependent, many of the similarities may be controlled, not necessarily by depth, but by distance from the land. This comparison may not therefore be valid.

## GENERAL DISCUSSION OF PELAGIC DEPOSITS

Whether the deposits described above are deep water or not one very important point must be noted. Pelagic facies are often present in a similar position in other ancient geosynclinal sequences. This can be seen in Upper Palaeozoic sediments of the Variscan geosyncline which extended from southern Ireland and southwestern England eastwards through Germany. Devonian and Carboniferous rocks of these regions often consist of laminated black shales with ostracods and trilobites, radiolarian cherts, and red nodular cephalopod-bearing limestones not unlike the ammonitico rosso (Tucker, 1974). Goldring (1962) has produced an elegant study combining absolute age determinations with the thicknesses of different facies in various regions. This enables an estimation of sedimentation rates to be made. The data suggest that deposition slowed markedly when the Old Red Sandstone fluviatile sedimentation was followed by the 'bathyal' pelagic facies in the axis of the geosyncline. Sedimentation speeded up again at different times in different regions when the bathyal phase was succeeded by flysch of the Culm Series.

A further common characteristic of ancient early geosynclinal pelagic sediments is their association with volcanic activity (Garrison, 1974). This typically consists of diverse pillow lavas, spilites, basalts, and serpentinites. This association, generally termed the Ophiolitic Suite, is found in the Jurassic and Cretaceous rocks of Greece, the Apennines, and the Alps. It occurs in Upper Palaeozoic rocks of the Variscan geosyncline in southwest England and Germany, and is present in the Lower Palaeozoic Caledonian geosyncline, notably in the southern uplands of Scotland. A further notable example occurs associated with large-scale gravity tectonics in the Oman Mountains of Arabia (Wilson, 1969). There has been considerable speculation that sub-marine volcanic eruptions locally increased the silica content of sea water, encouraging population explosions of radiolaria. Aubouin (1965) has shown, however, that radiolarian chert formation was independent of vulcanicity in the Jurassic and Lower Cretaceous sediments of the Pindus furrow in Greece. It might be interesting to relate the present distributions of radiolarian oozes to sub-marine vulcanism.

Finally one more example of possible ancient deep-water sediments deserves mention. This occurs in the island of Portuguese Timor between Australia and southeast Asia. Here some curious beds of Upper Cretaceous age lie beneath a thick Tertiary flysch sequence. In western Timor these contain red clays with manganese nodules, sharks teeth, and rare fish bones. In eastern Timor there is a diverse assemblage of radiolarian cherts, calcilutites, marls, turbidites, and shales rich in iron and manganese and containing manganese nodules. These sediments have been the subject of a detailed study by Audley-Charles (1965; 1966). The nodules of western Timor are geochemically similar to Recent deep-sea nodules while the eastern Timor examples show characteristics midway between Recent deep-sea and shelf nodules. Here again, however, the direct correlation of depth and geochemistry has been questioned by Price (1967) who, as previously mentioned (p. 288), has suggested that the chemical variation of modern manganese nodules is controlled by oceanicity rather than depth.

To conclude this discussion of ancient pelagic sediments the following points seem to be important. Turbidite flysch sequences often overlie a distinctive facies which can be recognized in various parts of the world in rocks of varying ages. This is composed of fine-grained limestones, shales, and marls. The limestones are sometimes red and nodular. They are associated with radiolarian cherts and, sometimes, with sub-marine vulcanicity. The fauna of this facies is typically pelagic; benthonic fossils are generally extremely rare. Sedimentary structures suggesting strong current action are lacking. Sequences of this facies generally show few stratigraphic breaks and are thinner than nearby time-equivalent facies.

It is clear that such rocks were slowly deposited in marine environments below the photic zone and away from strong current action. In many ways these sediments are comparable to Recent deep-sea sands. Is this similarity due to the fact that they both formed in deep water, or due to distance from the land?

## ECONOMIC SIGNIFICANCE OF DEEP-SEA DEPOSITS

Pelagic deposits are of interest for a number of reasons. They may be source beds from which petroleum may be expelled. They also host several types of metallic ore bodies. The oceans are generally oxygen deficient between about 200–1500 m (Emery, 1963). The upper limit corresponds to the average depth of the continental shelf margin. Thus barred basins whose floors lie between 200–1500 m may be areas where organic rich sediments can accumulate and ultimately become petroleum source beds. Detailed analysis of recent D.S.D.P. data has shown that there is an optimum sedimentation rate for the preservation of organic matter. This optimum rate varies with sediment type (Johnson-Ibach, 1982). It is about 14 m/m.y. for calcareous muds, 21 m/m.y. for siliceous oozes, and about 40 m/m.y. for clays. Clays are the sediments with the highest preserved organic content. Johnson-Ibach suggests that when sedimentation rates are too slow then organic matter is oxidized, when they are too high the organic content is diluted.

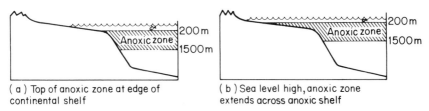

(a) Top of anoxic zone at edge of continental shelf

(b) Sea level high, anoxic zone extends across anoxic shelf

*Figure 11.4* Cross-sections to show how the oceanic anoxic zone may extend across continental shelves during periods of high sea level. This will favour the deposition of transgressive petroleum source beds.

At the present day the optimum situation for organic rich sedimentation lies below the present depth of the continental shelf. At times of globally high sea level however, the anoxic zone may extend across the continental shelf (Fig. 11.4). This may explain the occasional occurrence of widespread organic rich petroleum source beds, such as those of the Upper Jurassic and Lower Cretaceous (Schlanger and Cita, 1982).

The pelagic environment also favours the development of some mineral deposits, notably manganese and sulphide ores (Cronan, 1980). The occurrence of manganese nodules and areas of sea floor encrustations has been already noted. Feasability studies have been carried out on the commercial exploitation of these deposits. Nodules recovered from the floor of the Pacific Ocean contain up to 20% manganese and traces of copper, nickel and other metals. There are several ancient bedded manganese deposits which appear to have formed in ancient deep sea environments, notably in Cyprus and Oman (see Hutchison, 1983, p. 26 for a review). These deposits generally overly pillow lavas and/or ophiolites and are interbedded and overlain by cherts. These are in turn succeeded by pelagic fine-grained limestones or turbidites. The sequence appears to indicate a gradual shallowing and increase in sedimentation rate from the time of formation of the manganese beds. The igneous rocks below the ore were formed as old ocean bed, on which the manganese nodules and encrustations

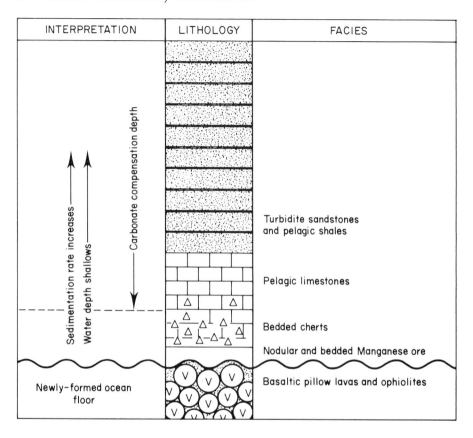

| INTERPRETATION | LITHOLOGY | FACIES |
|---|---|---|

*Figure 11.5* Figure to show the lithology, facies and environments of ancient deep marine manganese ores and associated deposits, such as those of Cyprus and Oman.

developed. The cherts formed as siliceous oozes, and the pelagic micrites were deposited in shallower conditions above the carbonate compensation level (Fig. 11.5).

Studies of recent mid-oceanic ridges, notably in the Red Sea, have revealed 'hot holes' where hot metal-rich brine springs emerge onto the sea floor (Degens and Ross, 1969; Rona 1984). These attain temperatures of 50–60 C. and salinities of over 200 p/p thousand. Hydrothermal sulphide ores are forming on the sea floor near the springs, aided by specialized thermaphilic bacteria. These recent deposits include sulphides of iron, copper, lead and zinc, together with traces of native copper, silver and gold. Ancient analogues of these deposits include the sulphide ore bodies of Rammelsberg, Germany; and Semail, Oman (see Jensen and Bateman, 1981 Chapter 8, for a review).

Some bedded barite also seem to have formed in deep marine conditions. The Monitor Valley deposits of Nevada are cited as an example. These occur in an Ordovician geosynclinal sequence of radiolarian cherts and shales. Barytes is interbedded with these and forms thin conglomerates mixed with phosphate nodules. The presence of re-worked barytes fragments strongly supports its primary origin (Shawe, Poole, and Brobst, 1969).

# REFERENCES

The description of Alpine deep-sea deposits was based on:

Aubouin, J., 1965. *Geosynclines.* Elsevier, Amsterdam. p. 335.
Barrett, T. J., 1982. Stratigraphy and Sedimentology of Jurassic bedded chert overlying ophiolites in the North Appennines, Italy. *Sedimentology*, **29**, 353–73.
Bernoulli, D. and Jenkyns, H. C. 1974. Alpine, Mediterranean, and Central Atlantic Mesozoic facies in relation to the early evolution of the Tethys. In: *Modern and Ancient Geosynclinal Sedimentation.* (Eds R. E. Dott, and R. H. Shaver). Spec. Pub. Soc. Econ. Pal. Min. No. 19. Tulsa. 129–60.
Farinacci, A. and Elmi, S., 1981. Rosso Ammonitico Symposium *Proceedings. Ed. Tec. Roma.* p. 229.
Garrison, R. E., and Fischer, A. G., 1969. Deep water limestones and radiolarites of the Alpine Jurassic. In: Depositional environments in carbonate rocks. (Ed. G. M. Friedman) *Soc. Econ. Min. Pal.*, Sp. Pub., No. 14, pp. 20–56.
——, 1974. Radiolarian cherts, pelagic limestones and igneous rocks in eugeosynclinal assemblages. In: *Pelagic sediments: on land and sea.* (Eds K. J. Hsu and H. C. Jenkyns) Blackwell Scientific Publications, Oxford. pp. 367–99.
Hallam, A., 1967. Sedimentology and palaeographic significance of certain red limestones and associated beds in the Lias of the Alpine regions. *Scott. J. Geol.*, **3**, 195–220.
Hudson, J. D., and Jenkyns, H. C., 1969. Conglomerates in the Adnet Limestones of Adnet (Austria) and the origin of 'Scheck'. *N. Jb. Geol. Palaeont. Mh.*, Jg. H.9, pp. 552–8.
Jenkyns, H. C., 1974. Origin of red nodular limestones (Ammonitico Rosso, Knollenkalke) in the Mediterranean Jurassic: a diagenetic model. In: *Pelagic sediments: on land and sea.* (Eds K. J. Hsu and H. C. Jenkyns) Blackwell Scientific Publications, Oxford, pp. 249–71.
Nocera, D. S. and Scandone, P., 1977. Triassic nannoplankton limestones of deep basinal origin in the central Mediterranean region. *Palaeogeog. Palaeoclim. Palaeoecol.*, **21**, 101–11.


Other references cited in this chapter:

Arrhenius, G., 1963. Pelagic sediments. In: *The Sea*, vol. III. (Ed. M. N. Hill) Interscience, N.Y. pp. 655–727.
Audley-Charles, M. G., 1965. A geochemical study of Cretaceous ferromanganiferous sedimentary rocks from Timor. *Geochem. Cosmochim. Acta*, **29**, 1153–73.
——, 1966. Mesozoic palaeography of Australasia. *Palaeogeography, Palaeoclimatology, Palaeoecology*, **2**, 1–25.
Berner, R. A., 1965. Activity coefficients of bicarbonate, carbonate and calcium ions in sea water. *Geochem. Cosmochim. Acta*, **29**, 947–65.
Cook, H. E., and Enos, P. (Eds), 1977. Deep water carbonate environments. *Soc. Econ. Pal. Min.*, Sp. Pub., No. 25. p. 325.
Cronan, D. S., 1980. *Underwater Minerals.* Academic Press, London. p. 362.

Degens, E. T., and Ross, D. A. (Eds), 1969. *Hot Brines and Recent Heavy Metal Deposits in the Red Sea.* Springer-Verlag, Berlin, p. 600.

Fischer, A. G., and Garrison, R. E., 1967. Carbonate lithifaction on the sea floor. *J. Geol.*, **75**, 488–96.

Friedman, G. M., 1965. Occurrence and stability relationships of aragonite, high-magnesium calcite and low-manganesium calcite under deep sea conditions. *Bull. Geol. Soc. Am.*, **76**, pp. 1191–5.

Goldring, R., 1962. The Bathyal lull: Upper Devonian and Lower Carboniferous sedimentation in the Variscan geosyncline. In: *Some Aspects of the Variscan Fold Belt.* (Ed. K. Coe) Exeter Univ. Press. pp. 75–91.

Hsu, K. J., and Jenkyns, H. C. (Eds), 1974. Pelagic sediments: on land and under the sea. *Internat. Assn. Sedimentol.* Spec. Pub., No. 1. Blackwell Scientific Publications, Oxford, p. 447.

Hudson, J. D., 1967. Speculations on the depth relations of calcium carbonate solution in recent and ancient seas. In: Depth Indicators in Marine Sedimentary Environments. (Ed. A. Hallam) *Marine Geology* Sp. Issue, **5**, No. 5/6, pp. 473–80.

Hutchison, C. S., 1983. *Economic Deposits and their Tectonic Setting.* Macmillan. London. p. 365.

Inderbitzen, A. L. (Ed.), 1974. *Deep-sea sediments.* Plenum Press, New York. p. 497.

Jensen, M. L. and Bateman, A. M., 1981. *Economic Mineral Deposits.* 3rd Edn. J. Wiley and Sons, Chichester. p. 593.

Lisitzin, A. P., 1972. Sedimentation in the world ocean. *Soc. Econ. Pal. Min.* Sp. Pub., No. 17. p. 218.

Lucas, G., 1973. Deep sea carbonate facies. *Bull. Centre Rech. Pau.*, **7**, 193–206.

Morris, D. A., 1976. Organic diagenesis of Miocene sediments from Site 341, Voring Plateau, Norway. In: Talwani, M., Udintsev, G., *et al., Initial Rept. Deep Sea Drilling Project.* **38**, Washington (U.S. Govt. Printing Office), pp. 809–14.

Price, N. B., 1967. Some geochemical observations on manganese-iron oxide nodules from different depth environments. In: Depth Indicators in Marine Sedimentary Environments. (Ed. A. Hallam) *Marine Geology*, Sp. Issue, **5**, No. 5/6, pp. 511–38.

Riedel, W. R., 1963. The preserved record: palaeontology of pelagic sediments. In: *The Sea*, vol. III. (Ed. M. N. Hill) Interscience, N.Y. pp. 866–87.

Rona, G. P., 1984. Hydrothermal mineralization and sea floor spreading centres. *Earth Sci. Rev.*, **20**, 1–40.

Ross, D. A., 1982. *Introduction to Oceanography.* Prentice-Hall, New Jersey. p. 528.

Schlanger, S. O. and Cita, M. B., 1982. *Nature and Origin of Cretaceous Carbon-Rich Facies.* Academic Press, London. p. 250.

Scrutton, R. A. and Talwani, M., 1982. *The Ocean Floor.* J. Wiley, Chichester. p. 332.

Shawe, D. R., Poole, F. G., and Brobst, D. A., 1969. Newly discovered bedded Barite deposits in East Northumberland Canyon, Nye County, Nevada. *Econ. Geol.*, **64**, 245–54.

Shepard, F. P., 1963. *Submarine Geology.* Harper and Row, N.Y. 557 p.

Stow, A. V., and Piper, D. J. W., 1984. *Fine-grained Sediments: Deep-Water Processes and Facies.* Blackwell, Oxford. p. 664.

Tucker, M. E., 1974. Sedimentology of Palaeozoic pelagic limestones: the Devonian Griotte (Southern France) and Cephalopodenkalk (Germany). In: *Pelagic Sediments: on land and under the sea.* (Eds K. J. Hsu and H. C. Jenkyns). Sp. Pub. Internat. Assn. Sedol. I. 71–92.

Wilson, H. H., 1969. Late Cretaceous and Eugeosynclinal sedimentation, gravity tectonics and ophiolite emplacement in Oman Mountains, southeast Arabia. *Bull. Amer. Assoc. Petrol. Geol.*, **53**, 626–71.

# 12   Conclusions

## SEDIMENTARY MODELS: MYTHICAL AND MATHEMATICAL

It should be apparent from the preceding chapters that environmental analysis is at best an imprecise art, rather than a deterministic scientific discipline. This is due to the diverse processes which control a depositional environment and hence the extremely variable nature of the resultant sedimentary facies.

One can never actually prove the depositional environment of a rock as one can derive a mathematical formula or repeat a physical or chemical experiment. We can describe the facies (effect, result), we can only guess at the environment (cause, process).

There are basically two possible philosophic approaches to the environmental analysis of sediments and the utilization of this knowledge. Consider the following simple problem: two boreholes, a few miles apart, pierced a time-equivalent carbonate formation. In one well the rock is an inter-bedded microcrystalline dolomite:pellet limestone facies. The corresponding interval in the second well is an argillaceous calcilutite with small pelagic foraminifera. Question: what is the nature of the transition between these two facies?

The first solution to this problem might use the following line of reasoning: the dolomite:pellet limestone facies is comparable to the deposits of Recent sabkhas and lagoons and the calcilutite facies is analogous to Recent open marine sediments forming below wave base. Between these two environments it is common today to find a high-energy environment of skeletal or oolite sand shoals; alternatively a reef might be present. The occurence of a similar facies is predicted between the two wells in question.

The second way of solving this particular problem might run like this: of 100 analogous situations on record, 90 have a reef or calcarenite facies separating the two already discovered. There is, therefore, a probability of 0·90 that the same will hold true for this particular case.

Both approaches may produce the right answer, but while the first is deductive, based on subjective understanding of Recent environments and deposits, the second is empirical and based on a mathematical analysis of objective criteria. The first approach is typical of geologists (as this book illustrates), while the second is more appropriate for a computer.

Both the geologist and the computer need some kind of conceptual framework to solve the problem. Thus the geologist has to have some idea of what Recent sabkhas, lagoons, shoals, and marine platforms are like. These mental pictures will be a function of his field experience, of the books and papers he has read, and of the prejudices and will-power of his teachers. The computer, on the other hand, will have no 'picture' of the environments, but a programmer will have previously fed it data which objectively classify all sedimentary facies and a large number of case histories. Even though the computer does not need to 'understand' recent processes to manipulate facies (results), both computer and geologist must have a logical basis for structuring and classifying the data. The data in the computer must have been analysed and programmed by a geologist liable to the same subjective prejudices as his less sophisticated counterpart.

In both cases, therefore, though the geologist has to interpret the data, which the computer does not, both have to have a conceptual framework of sedimentary facies.

### Sedimentary models: mythical

As discussed in the first chapter of this book (p. 3), many writers have considered the problem of classifying and defining sedimentary environments and the facies which they generate. This approach has led to the concept of the sedimentary model (Walker, 1979). This states that there are, and always have been, a number of sedimentary environments which deposit characteristic facies. These may be classified into various ideal sedimentary systems or models. This concept is fundamental to this book. The case histories in it have been selected as examples which most closely approximate to the writer's subjectively defined ideal models. In another book (Selley, 1976 and 1982) I attempted to state the reasoning behind the sedimentary model concept in the following way:

**Observation**: There are a finite number of sedimentary environments on the earth's surface today.

*Figure 12.1* Some kind of conceptual model is essential for any imaginative kind of interpretation. Anon. 1967. Reproduced from *Atlas* 1. No. 4, p. 62, by courtesy of Elsevier Publishing Company.

(*Qualifications*: some geopedant will always be able to show that no two similar environments are exactly the same. Secondly, boundaries between adjacent environments are often transitional).

**Observation**: The stratigraphic record contains a finite number of sedimentary facies that occur repeatedly in space and time.

(*Qualifications*: no two similar facies are exactly identical, and facies boundaries are often gradational.)

**Interpretation**: The parameters of ancient sedimentary facies can be matched with those of recent deposits of known environment. Thus may the depositional environment of ancient sedimentary facies be known.

**Conclusion**: There are, and always have been, a finite number of depositional environments which deposit characteristic facies; these may be classified into a number of ideal sedimentary models.

There is, however, no such thing in nature as an ideal or typical fluviatile environment or facies, just as the concept of the ideal cabbage exists only in the minds of judges of horticultural shows. Some sort of conceptual model is essential for any kind of imaginative interpretaion (Fig. 12.1).

## Sedimentary models: mathematical

Attempts have been made to define environments and facies mathematically. These projects are now becoming more and more feasible, using the abilities of computers to store and statistically manipulate large amounts of data.

An early attempt to define a sedimentary system mathematically was made by Sloss (1962) and amplified by Allen (1964). The formula proposed by Sloss is:

$$\text{Shape} = f(Q, R, D, M)$$

where $Q$ is the volume of the detritus supplied per unit time; $R$ is the rate of subsidence; $D$ is a dispersal factor of the rate of transport; $M$ is the composition of the material.

This is essentially a process : response model which expresses both the dynamics of the environment as well as the end result. Since we can only guess at the process, but observe its result, it may be safer, to begin with, to define a facies by its five parameters thus:

$$\text{Facies} = \Sigma(G, L, Ss, Pp, F)$$

where $G$ is geometry, $L$ is lithology, $Ss$ is sedimentary structures, $Pp$ is the palaeocurrent pattern, and $F$ is its palaeontology.

Both these formulae are insoluble until it is possible to numerate the parameters (how do you express a palaeocurrent pattern mathematically in a way which can be integrated with a mathematical statement of the associated fossils?).

This approach is, however, a useful starting point to consider ways of defining facies and diagnosing environments mathematically.

Consider first a very simple case. Ignoring geometry and palaeocurrents for the moment, imagine that a facies can be defined only by its chemical composition, grain size, and fauna, i.e.

$$\text{Facies} = \Sigma(C, G, F)$$

where $C$ is the chemical composition, $G$ is the depositional grain size, and $F$ is its palaeontology. Assume too that these three parameters can be calibrated on linear scales. Chemical composition ranging, say, from silica, through calcium carbonate to evaporite minerals; grain size ranging from conglomerate to clay, and fossils being interpreted on a scale which reads from marine through brackish and freshwater to terrestrial.

On this basis it is possible to define a given facies in three-dimensional space using these three variables as mutually perpendicular vector axes (Fig. 12.2).

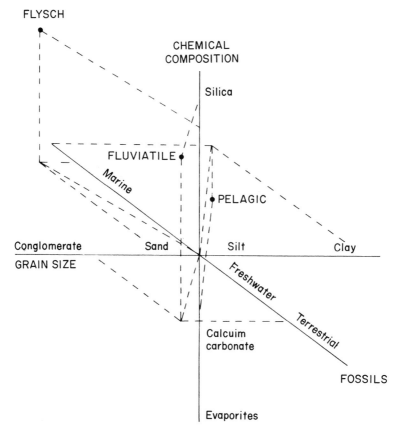

*Figure 12.2* Facies fixed in three-dimensional space using their chemistry, grain size and palaeontology as vectors.

This is an extremely naïve way of structuring a very complex problem, but it points the way to a more sophisticated approach. A facies is the sum of many more variables than can be expressed graphically in three-dimensional space. It is, however, possible to position objects defined by many variables in multi-dimensional space using the statistical technique of factor analysis (Harbaugh and Merriam, 1968, p. 179). The human mind is incapable of envisaging more than four dimensions (the fourth, time, is not applicable to this situation), but this can be done by a computer.

If one can numerate all the various parameters of a facies, it is then possible to use factor analysis to fix its position in multidimensional space. This would

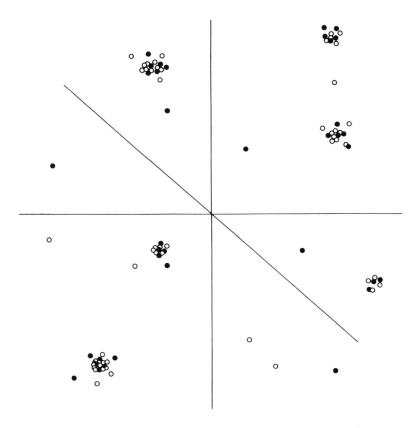

*Figure 12.3* Three-dimensional representation for facies fixed in multi-dimensional space using factor analysis. Black spheres indicate location of ancient facies, white spheres indicate Recent environment case histories. Clusters of spheres indicate location of various sedimentary models (e.g. fluviatile, reef, etc.). Randomly distributed spheres indicate case histories transitional between commonly occurring types.

prove or disprove the concept of the sedimentary model. If this idea is valid, one would predict clusters in space of fluviatile facies, reefs, and so on. These clusters should be composed of both Recent and ancient case histories. Each cluster would represent an 'ideal' model, though a hole in the centre of a cluster would mean that a 'type' specimen was yet to be found (if it exists). Isolated case histories scattered through space between clusters would represent transitional environments (Fig. 12.3).

I produced the foregoing geofantasy when I wrote the first edition of this book in 1969. This geofantasy has now become a routine technique in facies analysis. The breakthrough came in a paper published by Rider and Laurier in 1979. These authors studied a deltaic petroleum reservoir (Fig. 12.4). In so doing they produced a sedimentary model in the form of a conventional 'ideal cycle'. They went further however. A normal technique of geophysical log analysis is to cross-plot data from different tools at the same depth. Normally it is the porosity logs that are used in this way to obtain accurate porosity values, and to identify the lithology of the reservoir. Rider and Laurier used this technique in a different way. They took a single genetic increment through the delta and, instead of only plotting data for the reservoir sands, they plotted data throughout the whole cycle (Fig. 12.5). Here was a way to define a sequence, not as a geologists geospasm, but with an objective set of data. Using the same technique it is of course possible to cross-plot readings throughout a single

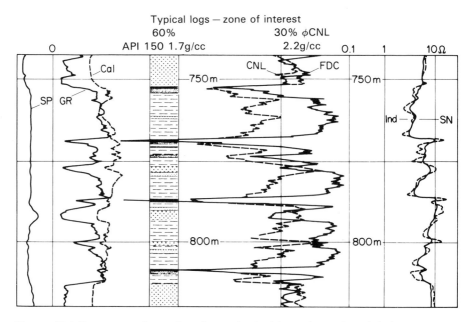

*Figure 12.4* Representative suite of geophysical logs through a deltaic sequence. From Rider and Laurier (1979).

borehole, or many (Fig. 12.6). Thus one can display a characteristic 'signature' for that particular sedimentary facies (Fig. 12.7). This technique may then be used to make objective comparisons with other facies, QED.

When a borehole is drilled however a considerable number of geophysical logs are often run. These may record formation resistivity at several distances from the well bore, an acoustic log, and several different types of radioactivity log. Thus at any depth interval up to ten different petrophysical properties of the formations may be recorded. Rider and Laurier plotted only two of these. Fig. 12.8 shows how facies may be displayed in three-dimensional space using the readings from three geophysical logs (cf. Fig. 12.2). Serra and Abbott (1980) have shown how this approach can be extended into multidimensional space. They did this using the well-established statistical technique of multivariate analysis. This has been widely employed in palaeontological and other classificatory schemes. Multivariate analysis can handle clusters of data in multidimensional space (Harbaugh and Merriam, 1968). Geophysical logs

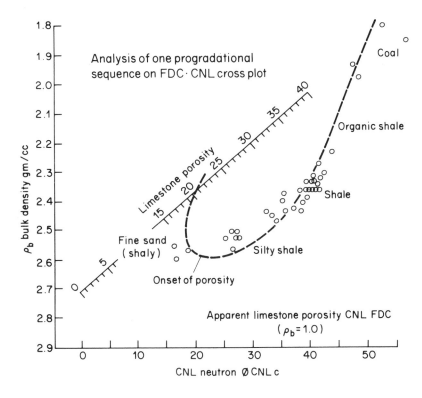

*Figure 12.5* Cross-plot of Bulk Density and Neutron log readings for a single genetic increment of the deltaic deposits illustrated in Fig. 12.4. From Rider and Laurier (1979).

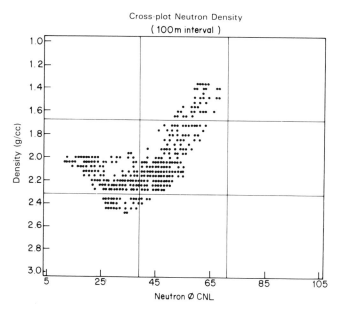

*Figure 12.6* Density-Neutron log cross-plot through 100 metres of the deltaic sediments characterized in Fig. 12.4. This objectively defines this particular sedimentary facies. From Rider and Laurier (1979).

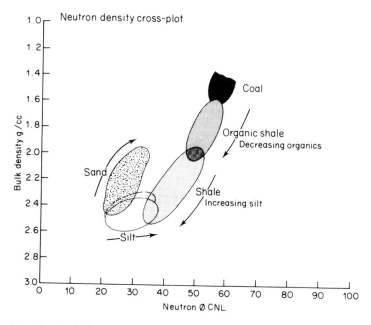

*Figure 12.7* Geological interpretation of the cross-plot illustrated in Fig. 12.6. From Rider and Laurier (1979).

provide the raw data that can be used in the cluster analysis of sedimentary facies. Serra and Abbot show how sediments can be objectively characterized by electrofacies which they define as: 'the set of log responses that characterizes a sediment and permits the sediment to be distinguished from others'. The end result of this multivariate analysis is the production of an electrofacies log. The scale of this indicates the cluster number. Cluster numbers have no sequential value (that is to say 4 is no larger than 3. It is rather like comparing apples, pears and bananas). Electrofacies logs enable sediments to be objectively defined and compared without the use of subjectively geologically defined facies. Where will it all end?

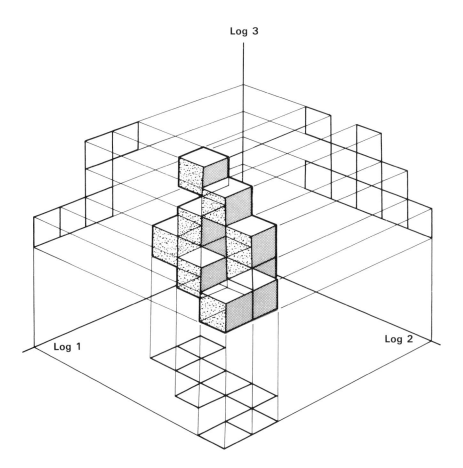

*Figure 12.8* Three-dimensional display of 'electrofacies' defined by three geophysical well logs. From Serra and Abbott (1980).

## REFERENCES

Allen, P., 1964. Sedimentological models. *J. Sediment. Petrol.*, **34**, pp. 289–93.

Harbaugh, J. W., and Merriam, D. F., 1968. *Computer Applications in Stratigraphic Analysis.* J. Wiley, N.Y. p. 282.

Moore, P. G., 1967. The Use of Geological Models in Prospecting for Stratigraphic Traps. *Proc. Seventh World Petrol. Cong.*, **2**, Elsevier, Amsterdam, pp. 481–6.

Rider, M. H. and Laurier, D., 1979. Sedimentology using a computer treatment of well logs. *Trans. Soc. Prof. Well Log Analysts.* 6th European symposium. London. Paper J. p. 13.

Selley, R. C., 1976. *Introduction to Sedimentology.* (2nd Edn., 1982). Academic Press, London. p. 426.

Serra, O. and Abbott, H. T., 1980. The contribution of logging data to sedimentology and stratigraphy. *Soc. Pet. Eng. Paper 9270.* p. 19.

Sloss, L. L., 1962. Stratigraphic models in exploration. *J. Sediment. Petrol.*, **32**, pp. 415–22.

Walker, R. G., 1979. *Facies Models.* Geoscience Canada Reprint Series. No. 1. Toronto. p. 211.

# Author index

Quanheng, Z., 111, 112

Rachocki, A. H., 44, 80
Ragsdale, J. A., 115, 143
Rainwater, E. H., 148, 162, 168
Ramsbottom, W. H. C., 128, 144
Randy Ray, R., 111, 113
Randohr, P., 242
Raudkivi, A. J., 13, 40
Rayner, D. H., 118, 143
Read, J. F., 194, 209
Read, W. A., 10, 40, 128, 144
Reading, H. G., 126, 143, 144
Rech-Frollo, M., 249, 268, 281
Reeckmann, A., 194, 204, 209, 233, 234, 244
Reeves, C. C., 102, 105, 113
Reidel, W. R., 283, 294
Reineck, H. E., 13, 40
Rezak, R., 232, 243
Richardson, E. V., 14, 41
Rider, M., 300, 304
Rider, R. T., 113
Riecke, H., 233, 242
Rigby, J. K., 241
Roberts, J., 243
Robertson, J. O., 233, 244
Rodriguez, J., 18, 40
Rognon, P., 208
Rohrlich, V., 12, 40
Rona, G. P., 292, 294
Ross, D. A., 292, 294
Rusnak, G. A., 249, 281
Rust, B. R., 44, 80, 81

Said, R., 205, 209
Sanders, J. E., 100
Sanford, R. M., 68, 80
Sangree, J. B., 41
Savage, R. J. G., 182
Scandone, P., 293
Scholle, P. A., 204, 209
Schlanger, S. O., 291, 294
Schleisher, J. A., 41
Schmidt, V., 208
Schramm, M. W., 73, 80
Schumm, S. A., 42, 80
Schwarzacher, W., 5, 40
Scott, J., 242
Scott, K. M., 254, 281
Scrutton, R. A., 283, 294
Seilacher, A., 18, 41, 203, 209

Selley, R. C., 2, 5, 10, 13, 15, 16, 18, 19, 41, 64, 77, 80, 134, 144, 168, 182, 202, 209, 271, 279, 281, 296, 304
Sellwood, B. W., 143
Serra, O., 301, 304
Shantser, E. V., 42, 80
Sharma, G. D., 184, 209
Shaver, R. H., 143
Shaw, H. F., 12, 41
Shawe, D. R., 293, 294
Shearman, D. J., 188, 204, 210, 233, 244
Shelton, J. W., 3, 41, 165, 168, 269, 282
Shepard, P. E., 17, 41, 117, 128, 143, 145, 212, 240, 250, 252, 253, 282, 284, 294
Sherrif, R. F., 41
Shimp, N. F., 12, 40, 41
Shinn, E. A., 188, 209
Shirley, M. L., 115, 143
Shotton, F. W., 92, 95, 100
Siemers, E. T., 165
Siever, P. E., 25, 40
Silver, B. A., 241
Simons, D. B., 14, 41
Simons, J. A., 128, 144
Singh, I. B., 13, 40
Sloss, L. L., 3, 39, 234, 244, 297, 304
Smalley, I. J., 93, 101
Smit, D. E., 168
Smith, D. G., 44, 80
Smith, N. D., 73, 80
Stamp, L. D., 159, 168
Stanley, D. J., 184, 210, 250, 267, 271, 279, 280, 282
Stanley, K. O., 112
Stanley, K. P., 89, 99
Stanton, R. J., 17, 38
Stark, P. H., 68, 79
Stauffer, P. H., 250, 282
Steers, J. A., 146, 168
Sterrett, T. S., 174, 183
Stewart, A. D., 52, 77
Stephens, R. W., 142, 144
Stockman, K. W., 39, 240
Stokes, W. L., 63, 81, 94, 99
Stout, J. L., 239, 244
Stow, A. V., 283, 294
Stride, A. H., 186, 210
Sturt, B. A., 249, 282
Sullwold, H. H., 269, 282
Summer, G. E., 207
Summerhayes, C. P., 210
Surdam, R. C., 112

# Subject index